T0202980

Communications
in Computer and Information Science

2122

Editorial Board Members

Joaquim Filipe ⓘ, *Polytechnic Institute of Setúbal, Setúbal, Portugal*
Ashish Ghosh ⓘ, *Indian Statistical Institute, Kolkata, India*
Lizhu Zhou, *Tsinghua University, Beijing, China*

Rationale

The CCIS series is devoted to the publication of proceedings of computer science conferences. Its aim is to efficiently disseminate original research results in informatics in printed and electronic form. While the focus is on publication of peer-reviewed full papers presenting mature work, inclusion of reviewed short papers reporting on work in progress is welcome, too. Besides globally relevant meetings with internationally representative program committees guaranteeing a strict peer-reviewing and paper selection process, conferences run by societies or of high regional or national relevance are also considered for publication.

Topics

The topical scope of CCIS spans the entire spectrum of informatics ranging from foundational topics in the theory of computing to information and communications science and technology and a broad variety of interdisciplinary application fields.

Information for Volume Editors and Authors

Publication in CCIS is free of charge. No royalties are paid, however, we offer registered conference participants temporary free access to the online version of the conference proceedings on SpringerLink (http://link.springer.com) by means of an http referrer from the conference website and/or a number of complimentary printed copies, as specified in the official acceptance email of the event.

CCIS proceedings can be published in time for distribution at conferences or as post-proceedings, and delivered in the form of printed books and/or electronically as USBs and/or e-content licenses for accessing proceedings at SpringerLink. Furthermore, CCIS proceedings are included in the CCIS electronic book series hosted in the SpringerLink digital library at http://link.springer.com/bookseries/7899. Conferences publishing in CCIS are allowed to use Online Conference Service (OCS) for managing the whole proceedings lifecycle (from submission and reviewing to preparing for publication) free of charge.

Publication process

The language of publication is exclusively English. Authors publishing in CCIS have to sign the Springer CCIS copyright transfer form, however, they are free to use their material published in CCIS for substantially changed, more elaborate subsequent publications elsewhere. For the preparation of the camera-ready papers/files, authors have to strictly adhere to the Springer CCIS Authors' Instructions and are strongly encouraged to use the CCIS LaTeX style files or templates.

Abstracting/Indexing

CCIS is abstracted/indexed in DBLP, Google Scholar, EI-Compendex, Mathematical Reviews, SCImago, Scopus. CCIS volumes are also submitted for the inclusion in ISI Proceedings.

How to start

To start the evaluation of your proposal for inclusion in the CCIS series, please send an e-mail to ccis@springer.com.

S. Satheeskumaran · Yudong Zhang ·
Valentina Emilia Balas · Tzung-pei Hong ·
Danilo Pelusi
Editors

Intelligent Computing for Sustainable Development

First International Conference, ICICSD 2023
Hyderabad, India, August 25–26, 2023
Revised Selected Papers, Part II

 Springer

Editors
S. Satheeskumaran 🆔
Anurag University
Hyderabad, Telangana, India

Valentina Emilia Balas
Aurel Vlaicu University of Arad
Arad, Romania

Danilo Pelusi
University of Teramo
Teramo, Italy

Yudong Zhang 🆔
University of Leicester
Leicester, UK

Tzung-pei Hong 🆔
National University of Kaohsiung
Kaohsiung, Taiwan

ISSN 1865-0929   ISSN 1865-0937 (electronic)
Communications in Computer and Information Science
ISBN 978-3-031-61297-8   ISBN 978-3-031-61298-5 (eBook)
https://doi.org/10.1007/978-3-031-61298-5

© The Editor(s) (if applicable) and The Author(s), under exclusive license
to Springer Nature Switzerland AG 2024

This work is subject to copyright. All rights are solely and exclusively licensed by the Publisher, whether the whole or part of the material is concerned, specifically the rights of translation, reprinting, reuse of illustrations, recitation, broadcasting, reproduction on microfilms or in any other physical way, and transmission or information storage and retrieval, electronic adaptation, computer software, or by similar or dissimilar methodology now known or hereafter developed.
The use of general descriptive names, registered names, trademarks, service marks, etc. in this publication does not imply, even in the absence of a specific statement, that such names are exempt from the relevant protective laws and regulations and therefore free for general use.
The publisher, the authors and the editors are safe to assume that the advice and information in this book are believed to be true and accurate at the date of publication. Neither the publisher nor the authors or the editors give a warranty, expressed or implied, with respect to the material contained herein or for any errors or omissions that may have been made. The publisher remains neutral with regard to jurisdictional claims in published maps and institutional affiliations.

This Springer imprint is published by the registered company Springer Nature Switzerland AG
The registered company address is: Gewerbestrasse 11, 6330 Cham, Switzerland

If disposing of this product, please recycle the paper.

Preface

The first International Conference on Intelligent Computing for Sustainable Development (ICICSD 2023) was held at Anurag University, Hyderabad, India, during August 25–26, 2023. This conference served as a key platform for the exchange of knowledge among academicians, scientists, researchers, and industry experts worldwide, focusing on pivotal areas such as digital healthcare, renewable energy, smart cities, digital farming, and autonomous systems. It aimed to facilitate the dissemination of innovative ideas and research findings in the realm of intelligent computing and its diverse applications. Recognizing the significant potential for advancement in these domains, a series of conferences have been planned to foster ongoing research for the betterment of society.

A total of 138 full papers underwent a rigorous review process. Each submission received a single-blind review by three subject matter experts with national and international recognition. Program committee members and reviewers actively participated in the peer-review process. Based on the reviews, the Program Committee chairs accepted 46 high-quality papers for presentation, resulting in a competitive acceptance rate of 33%. We were fortunate to host five distinguished keynote speakers: Yudong Zhang from the University of Leicester, UK; Fernando Ortiz-Rodriguez from Tamaulipas Autonomous University, Mexico; Sasikanth Adiraju from GE Power, Hyderabad, India; Manikanta Kumar from Hyundai Mobis, India; and Devarpita Sinha from Mathworks India Pvt. Ltd., India Their talks provided a unique opportunity for us to gain valuable insights from leaders in their respective fields.

We are grateful to Communications in Computer and Information Science (CCIS), Springer, for publishing the conference proceedings. We extend a special thanks to the Anurag University leadership team for providing the support to host this conference as an event at this institute. Their commitment made ICICSD 2023 a successful event for the institute. We extend our heartfelt appreciation to all the authors and co-authors who contributed their work to this conference, as well as to the Technical Program Committee members and reviewers for their invaluable expertise in selecting high-quality papers for inclusion. Their dedication and commitment have been instrumental in making ICICSD 2023 a resounding success.

August 2023

S. Satheeskumaran
Yudong Zhang
Valentina Emilia Balas
Tzung-pei Hong
Danilo Pelusi

Organization

Program Committee Chairs

S. Satheeskumaran	Anurag University, India
Yudong Zhang	University of Leicester, UK
Valentina Emilia Balas	Aurel Vlaicu University of Arad, Romania
Tzung-Pei Hong	National University of Kaohsiung, Taiwan
Danilo Pelusi	University of Teramo, Italy

Technical Program Committee

Yi Pan	Georgia State University, USA
Tamal Bose	University of Arizona, USA
Joy long-Zong Chen	Da-Yeh University, Taiwan
Eva Reka Keresztes	Budapest Business School, Hungary
Abzetdin Adamov	Qafqaz University, Azerbaijan
Raghav Katreepelli	Intel, USA
Arumugam Sundaram	Navajo Technical University, USA
Sarang Vijayan	Nova Systems Australia and New Zealand, Australia
Gajendranath Choudhary	IIT Hyderabad, India
S. Sridevi	Thiagarajar College of Engineering, India
Amrit Mukherjee	Jiangsu University, China
A. Senthil Kumar	Dayananda Sagar University, India
Magendran Koneti	Qualcomm, India
Jitendra Kumar Das	Kalinga Institute of Industrial Technology, India
K. Sasikala	Vels Institute of Science, Technology & Advanced Studies, India
Nitin Pandey	Amity University, India
Suresh Seetharaman	Sri Eshwar College of Engineering, India
S. Meenakshi Ammal	Pinter Fani Asia Pvt. Ltd., India
P. Chandrashekar	Osmania University, India
B. Rajendra Naik	Osmania University, India
A. Rajani	Jawaharlal Nehru Technological University, India
Chaudhuri Manoj Kumar Swain	Anurag University, India
Manoranjan Dash	Anurag University, India

Additional Reviewers

B. Subbulakshmi	Thiagarajar College of Engineering, India
Ch. Rajendra Prasad	SR University, India
C. Rajakumar	Vidya Jyothi Institute of Technology, India
Tejaswini Kar	Kalinga Institute of Industrial Technology, India
Maniknanda Kumar	New Horizon College of Engineering, India
G. Ananthi	Thiagarajar College of Engineering, India
K. V. Uma	Thiagarajar College of Engineering, India
M. Nirmala Devi	Thiagarajar College of Engineering, India
A. Bharathi	Renault Nissan Technology & Business Centre, India
Sasmita Pahadsingh	Kalinga Institute of Industrial Technology, India
S. Karthiga	Thiagarajar College of Engineering, India
S. Sasikala	Velammal College of Engineering and Technology, India
Tzung-Pei Hong	National University of Kaohsiung, Taiwan
Sambhudutta Nanda	Vellore Institute of Technology, India
Thangavel Murugan	United Arab Emirates University, UAE
Rushit Dave	Minnesota State University, India
Umesh Sahu	Manipal Institute of Technology, India
Sukant Sabut	Kalinga Institute of Industrial Technology, India
Giuseppe Aiello	University of Palermo, Italy

Organizing Committee

T. Anilkumar	Anurag University, India
N. Mangala Gouri	Anurag University, India
D. Haripriya	Anurag University, India
Rajesh Thumma	Anurag University, India
E. Srinivas	Anurag University, India
D. Krishna	Anurag University, India
M. Kiran Kumar	Anurag University, India
Kumar Neeraj	Anurag University, India
B. Srikanth Goud	Anurag University, India
P. Harish	Anurag University, India
G. M. Anitha Priyadarshini	Anurag University, India
B. Hemalatha	Anurag University, India
N. Sharath Babu	Anurag University, India
L. Praveen Kumar	Anurag University, India
S. Amrita	Anurag University, India

G. Anil Kumar   Anurag University, India
M. Kusuma Sri   Anurag University, India
J. Aparna Priya   Anurag University, India
G. Renuka   Anurag University, India
P. Lokeshwara Reddy   Anurag University, India

Contents – Part II

Contents – Part I

A Cognitive Architecture Based Conversation Agent Technology for Secure Communication

Preety[1]($\boxtimes$), Jagjit Singh[2], and Meenakshi Yadav[3]

[1] Christ University Delhi NCR, Ghaziabad, India
sunnypreety83@gmail.com
[2] Department of Artificial Intelligence and Data Science Koneru Lakshmaiah Education Foundation, Vaddeswaram, Andhra Pradesh, India
[3] Galgotias College of Engineering and Technology, Greater Noida, Uttar Pradesh, India

Abstract. This paper outlines a multi-agent system-based approach to provider selection. Suppliers in the supply chain are different and the demand and supply levels are high. Buy agents will find the right supply agent in our approach. First, the multi-layer classification system is used to rationally arrange and overall selection on suppliers and buyers. Secondly, the purchase information is organized by the supplier agent to improve device performance. The assessment process is then used to select the suppliers initially. In addition to selecting the correct provider and maximizing the value of the purchaser, the time negotiating mechanism is implemented.

Keywords: Agent Technology · Multi-Agent · Case Base Reasoning Systems · Decision Support System · JADE

1 Introduction

The selection of suppliers is a crucial aspect of control of the supply chain. This critical problem has been solved through numerous approaches. There have currently been several investigations into the choice of suppliers focused on multi-agents. The Information Agent and Function Agent were introduced to increase agent efficiency. The information agency provides financial data and the function agency performs the homologous function of negotiating in such a structure with other agents. Some previous papers introduced an agent-based method of seller selection and identified the agent's type and applied mechanisms for negotiating —Including negotiating goals, procedures, and ways of reasoning [1]. Berman and Yurin grouped agents function to increase the effectiveness of the multi-agent scheme using a fluffy framework to minimize the time to submit information [5]. Recent developments in the Upstart algorithm allow the selection of supplier problem to be addressed in supply chain [10]. Since the fuzzy logic approach has a very important feature in dealing with different circumstances, several researchers have suggested a fuzzy logic approach to provider selection. Cengiz Kahraman, UfukCebeci, and ZiyaUlukan have used the flexible method of analytical hierarchy (AHP) to pick the best supply company that meets the requirements established [2]. In [9], the dilemma

© The Author(s), under exclusive license to Springer Nature Switzerland AG 2024
S. Satheeskumaran et al. (Eds.): ICICSD 2023, CCIS 2122, pp. 1–12, 2024.
https://doi.org/10.1007/978-3-031-61298-5_1

of select multi-criteria vendors is defined as the weighted linear program. Both of the above help find the appropriate supplier.

There are some inconveniences, but in supply chain management, for example, several types of firms exist because it is not ideal to group the details of vendors and customers using individual-level classification tasks. In [5], in the sense of lower participation, object agents are placed at a level if the first collection fails, so there can be some risks for the buyer. If the risk level is not assessed, buyers' benefits can be affected. The role of the research paper written by Yun Peng's [10] in the negotiating process, the provider who fulfills the terms shall be identified. However, they could not maximize the value for the customer if several vendors comply with the terms. According to CBR approach a multilevel classification technique to address the issue at single levels [32].We recommend a way to group the buyer's information and then deliver the collective information to the suppliers, based on the fuzzy model suggested by [5]. The time of transmission of the message is then decreased, with more improvement on device performance.

This paper work also applies the risk degree for assessing the mechanism, calculating the anticipated utility, and determining with whom to collaborate. Timing mechanism used to increase the buyer's benefit based on the negotiating mechanism suggested in [10].The document is arranged accordingly. Section 2 describes the proposed mechanism for the selection of suppliers. Section 3 lists the intended reputation-based and risk-based utility assessment framework. Section 4 describes how the supplier may be selected and purchased. Section 5 provides a time-negotiation mechanism. Section 6 shows the simulation result. Last section concludes the paper.

2   Literature Review

In heterogeneous environments, agents operate in cooperation with other agents to fix complicated issues. Agent technology are keen in organizing individual agents' local conduct to provide a suitable conduct at the system level. Usage of intelligent agent gives capacity and system of scheme itself an even higher quantity of flexibility. Software development is becoming progressively hard with these new intricacies. It is therefore important that our procedures to build the inherently complicated distributed software that needs to operate in this setting are sufficient for the job. This section presents a methodology to design interacting agents for these systems.

Agents are autonomy capable of independently performing at least portion of their functionality and pursuing objectives independently. They are intelligent in the context that agents in one or even more areas of implementation have some specific understanding. The agents can obtain data or respond to their environment's circumstances. They are reactive so that they respond properly to their environment's inputs. They are focused on pro-activity and goal. Based on their prior experience, the agents alter their conduct. They are portable in the context that agents are transported from one network node to another. The key characteristics are communicative and cooperative agents.

Multi agent systems are those system that communicate with various agents to fix issues. MAS agents understand when and with whom to communicate. The integral circulation and complication of multi agent systems are common characteristics. Multi agent systems allotment and elastic nature lead to increased speed, reusability, robustness, and scalability. At Multi-Agent System,

- Every agents have partial data.
- The controlling system are devolved.
- The information are reorganized
- Asynchronous is computing

When developing MAS, several problems need to be resolved, such as when the agent how interact with their policy cooperate and compete to achieve its prototype goals effectively, agents take such time need to address sub-problem jointly, also under this situation agents would comprehend the capacities of other agents, how agents can decay their jobs and objectives (and allot sub-objectives and sub-jobs). What languages or procedures to add, how to represent agents and explain the behavior, plans and understanding of other agents so that they can communicate each other, how can agents indicate and explain their state relationships, how can agents be allowed acknowledge along with manage disputes in the middle of agent, in what ways to guarantee that multi-agent systems are properly defined, how to implement agents as the demand for stronger, more effective and more flexible agents increases, so fall the compression upon designers. At the same moment, building agent to accomplish different other duties would result in complex growth and increasing maintenance agents being usually intended for a particular purpose. If agents are required to undertake more duties, system can either improve their complexity (which improves the effort for growth), or it can make them function together. Coordination between different agents is imperative and effectual message is needed to be successful. For coordination involve a dual communication language medium. The media of language and communication are critical for agent-to-agent collaboration.

In this part, a multi-agent system-based supplier selection mechanism (MAS) proposed. Here, we define the frame as seen in Fig. 1 using the hierarchical structure [11]. Supply-agent delivers commodities in this system And the purchasing agent purchases commodities [12]. The agent with the complete understanding of the buying agent is called the buying agent, and the agent with the details of the supply agent is called the supply agent. Officer. Providers and consumers apply to their respective classes in accordance with their responsibilities. The service officer of either the supply service agents or buy-service agents supplies all knowledge., whose task is to receive data from the clients and to deliver it to the providers. All service agents are administered by the manager [13].

When purchasing data to find a suitable supplier, Supply-Service-Agent selects a series of mechanisms for efficient Collection, if the Stuffily-Agent is looking forward to cooperating, the buyer will start negotiating the time machine with the Supply-Agent [14].

In heterogeneous environments, agents operate in cooperation with other agents to fix complicated issues. MAS used to organizing individual agents' local conduct to provide a suitable conduct at the system level. Usage of intelligent agent gives capacity and system of scheme itself an even higher quantity of flexibility. Software development is becoming progressively hard with these new intricacies. It is therefore important that our procedures to build the inherently complicated distributed software that needs to operate in this setting are sufficient for the job. This section presents a methodology to design interacting agents for these systems. Agents are autonomy capable of independently performing at least portion of their functionality and pursuing objectives independently. They are intelligent in the context that agents in one or even more areas of implementation have some specific understanding. The agents can obtain data or respond to their environment's circumstances. They are reactive so that they respond properly to their environment's inputs. MAS agents understand when and with whom to communicate. When developing MAS, several problems need to be resolved, such as when the agent how interact with their policy cooperate and compete to achieve its prototype goals effectively, agents take such time need to address sub-problem jointly, and all sounder this situation agents would comprehend the capacities of other agents, how agents can decay their jobs and objectives (and allot sub-objectives and sub-jobs) [21]. The media of language and communication are critical for agent-to-agent collaboration [22]. Multidisciplinary fields came together to collaborate on the development of multi-agent systems in an effort to provide resilient, intelligent, and distributed applications. Unfortunately, there are many instruments concentrated a particular architecture of agents or the level of detail necessary to properly manage complex growth of systems. In the job, agents developed both a full lifecycle methodology and a complimentary environment to analyze, design and implement heterogeneous multi-agent systems. We are creating a multi-agent system engineering (MASE) approach. It's hard to build multi-agent systems. They have all the problems with traditional, simultaneous systems and additional challenges demands for flexibility and advanced interactions [23]. Object- oriented design has reached some maturity and offers a stable basis to build on. However, object-oriented research methods do not apply directly to agent systems-typical agents in both design and conduct are considerably more complicated than objects. A process consisting of multiple agents communicating with each another is a multi agent system. Overall, agents act for users with various aims and motivations. In most cases [24], to communicate effectively, we require cooperation, communication and negotiation as much as individuals can.

Reasoning on a case-based model (CBR) is more critical as relatively recent problem-solving strategies. However, there is still a tiny number of individuals with CBR theoretical or practical knowledge of first hand [29]. The primary purpose of this review is to provide individuals new to CBR with a thorough overview of the topic. The paper outlines CBR's growth in the United States in the 1980s. It will define CBR's basic methods and focus on strategy reasoning of model-based schemes. An critic analysis the CBR software instruments currently accessible is accompanied by descriptions of CBR apps from scholarly studies as well as three CBR technologies currently used commercial. Each of the three business cases analyses outlines characteristics that made CBR appropriate as implementation in particular. In addition, the last study based discusses a methodology of growth to implement CBR system. A study agenda document ends with CBR [28]. A complementary article provides a comprehensive classified bibliography of CBR studies. In the memory alone, reality takes shape, Marcel Proust. The one most successful stories of Artificial Intelligence (AI) study is expert or knowledge-based systems (KBS). More than 2000 KBS were discovered in commercial service by UK Department of Trade & Industry in a latest study (study excluded KBS in university study labs). The first documented KBS (the classical trinity) has been around 20 years since schemes: PROSPECTOR, DENDRAL and MYCIN was recorded, but there has been little change in KBS ' fundamental configuration in that moment. Today's systems and earlier KBS totally focused on an explicit knowledge prototype needed to fix a challenge, so-called the system of the second generation, using a deep causal model that allows a process to think by first principles [32]. Nonetheless, whether knowledge is shallow or deep, a specific domain model still needs to be constantly elicited and implemented in laws or later as object models. However, although model-based KBS is undoubtedly successful, developers of these schemes in many industries have encountered several issues [30]:

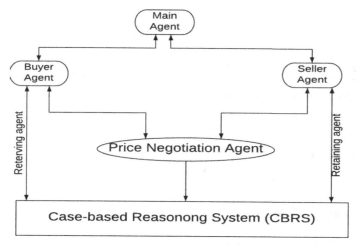

Fig. 1. Framework of Buyer and seller purchase system

3 Evaluation Based on Reputation and Risk Degree

In the supply chain structure, there are many vendors [15]. Some could be annoying attitudes towards unjust gains. The value of the purchaser would also be affected. Therefore, in supply chain management [31], the credibility graduate assessment process and risk assessment mechanism are of great importance [3, 8]. These systems provide highly reputable agents with more chances of collaboration, thus eliminating the anguishing conduct and encouraging fair conduct [16].

The following is the way to build reputation:

Ordered: the duration of the protocol order fulfill agreed: Ordered: RD protocol time: degree of credibility

The RD is of [0,1] range. The larger the ordering of the larger RD, i.e., the stronger the prestige, can be shown by Formula 1. Risk Rx is seen in the following conditional probabilities, if arranged with the suppliers:

$$Rx = \{(Sx1, Rx1), \dots, (Sxi, Rxi), \dots, (Sxn, Rxn)\}$$

$Sx_i(i=1,2,...,n)$: is a collection of different suppliers

$Rx_i(i=1,2,...,n)$: Risk size if Sx_i cooperates

The higher the supplier's prestige, the lower the chance the purchaser takes.

Therefore, they would make every effort to boost their prestige to improve their collaboration opportunities. To maximize the success level, customers work with vendors with lower risk [17]. This paper aims to demonstrate risk injury. The following formula sets out the expected utility:

Y refers to the cost of the products. α represents the level of risk; when cooperating with a provider it reflects the chance of risk threatening. P is the expression of logistics costs. By using the following formula, the estimated costs for cooperation with the supplier could:

$$Fi = (Y + P) * (1 - \alpha) + (Y) * \alpha$$

When you work with the supplier, Qi is the estimated cost. Cooperation succeeds, S is a gross profit.

Via the formula the benefit ab was calculated:

$$ab = S - Qi$$

And the impact is presented as expected The predicted utility decides to what extent suppliers and purchasers are required to cooperate.

4   Information Retrieval for Buyer and Suppler

To deliver purchasing information to the vendors correctly, when a procurement request from the buyer agent is sent, the The service agent must record all information from buying agents such as the source ip address of the purchaser agent, particulars and the sales order, etc. [18]. A Service-Agent can give information to the right vendors after the purchase information is registered [19].

Taking into consideration that there are maybe multiple purchase information's sent to the same provider and that communication of the message may be subject to the issue of network arrest, Therefore, before we forward purchasing information to the vendors, the management agent would have preferred to collect the purchase information A huge amount of time will be wasted if we deliver a message one at a time. It will be placed by the Service Agent category when purchasing material [20].

Given that the service agent cannot wait indefinitely for the purchasing information, the operation to group the purchasing In a certain period, details must be completed. The details collected will be sent after the time has elapsed. The concept is the same as above for organizing the buying results. Supply –Reputed degree for Agents. This reduces the time of transmitting messages and increases the overall machine performance. We will first discuss the initial list of vendors and later discuss the method of grouping the purchase details and submitting the collected purchasing information in depth to the chosen suppliers. The service agent picks and transfers the purchasing information to the appropriate agent depending on the prestige degree. How should I choose?

Calculate the average (A_1) worth of all at first

$$A_1 =^1 n_1 \sum i =^n 1 \, \text{Goodwill} \, Degree(i)$$

N_1: Represents number of the Supply-Agents required
A_1: the average of all Supply-Agents' goodwill degree
If I have goodwill higher than A_1, the service agent would answer (Fig. 2).

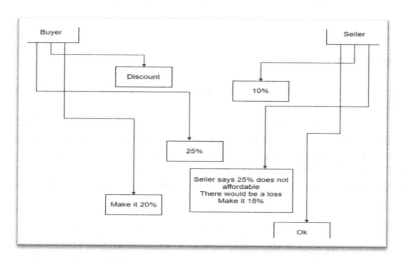

Fig. 2. Conversation Between Buyer and seller

The Service-Agent will send a note to the Purchase-Agency and inquire whether to send details on the purchase to a lower prestige level of the same company Supply-Agent. If the buyer agents agree to this, the purchasing details will be sent by the service agent to the respective supplier. It would also submit Buy-Agent information and ask if the purchasing information can be sent to other services providers. Thirdly, two need to be taken care of. While this approach is useful for purchase agents, more machine resources are consumed. Therefore, if the machine is busy, the third option will not be safer. Either way, Buy-Agent calculates the expected utility for negotiating with the Supply-Agent.

5 Intercession Process

An agent Communication Language (CL) is the accurate syntactic, semantic, and pragmatic language that forms the foundation of communication between autonomous software agents. Communication language (ACL) MAS functional agencies are using the common ACL for data transfer, knowledge sharing, and mutual negotiation [1]. And if customer are delighted with the new dealer, he/she do not have to choose the agent at once in our everyday economic behavior. Customer will begin choosing the happiest agent from the other agent. The negotiation process used for realizing this concept in this section. The process is used to find the most appropriate provider for maximum profit. Buy-Agent may negotiate with more than one supporter in this section. This process is used to find the most appropriate provider for maximum profit. Buy- Agent may negotiate with more than one supporter in a given period.

The time bargaining process is: When the appropriate supplier has been found, the Buyer-Agent may agree or connect the provider to the waiting options queue and the provider discuss more with other suppliers. When the time comes, the negotiations with all vendors will conclude. The purchasing agent would choose the most appropriate provider to achieve maximum value by compare the retailer representatives in the waiting line.

6   Negotiation Process

Multi-agent machine simulator software, such as JADE, SWARM, etc. There are plenty. Many investigators are currently engaged in simulating a multi-agent solution to the supplier selection problem [4, 6, 7]. Several simulations of the JADE platform have been carried out Validating the relevance of the processes proposed in this article. To explain this paper work product seller example used initially. A tables used that contains fifty sellers from s1 to s50 and each seller delivers information on how many days a product can be reached to destination and how much discount provided (Table 1).

Table 1. Table of seller and buyer

S No	Seller	Days	Discount	S No	Seller	Days	Discount
1	S1	2	3	26	S26	16	8.7
2	S2	3	4	27	S27	16	8.8
3	S3	5	4.2	28	S28	16	8.9
4	S4	6	5	29	S29	16	9
5	S5	7	5.2	30	S30	16	10
6	S6	8	5.3	31	S31	16	11.1
7	S7	9	5.4	32	S32	17	11.2
8	S8	9	5.6	33	S33	17	11.3
9	S9	9	5.8	34	S34	17	11.4
10	S10	10	5.9	35	S35	17	11.5
11	S11	10	6	36	S36	17	11.7
12	S12	10	6.1	37	S37	17	11.8
13	S13	10	6.2	38	S38	17	12
14	S14	11	6.3	39	S39	20	13
15	S15	12	6.4	40	S40	21	14
16	S16	13	6.6	41	S41	22	15
17	S17	14	6.7	42	S42	23	16
18	S18	14	6.8	43	S43	24	17
19	S19	14	6.9	44	S44	26	17.2
20	S20	14	7	45	S45	27	17.5
21	S21	14	8	46	S46	27	17.8
22	S22	15	8.1	47	S47	27	17.9
23	S23	16	8.2	48	S48	28	18
24	S24	16	8.4	49	S49	29	19
25	S25	16	8.5	50	S50	29	20

The calculation of the supplier number, number and price of items sponsored by the supplier, buyer number, number and price of the items bought by the buyer must be carried out during the simulation process. The other criteria such as when to commit to a protocol, when the protocol is performed, logistical costs, and supply agent credibility are all arbitrarily set in motion in a certain field as well. After setting the first parameters, the autonomous system agents communicated with each other according to the mechanism suggested in this article. The following two fingers are obtained from the outcome of the provider selection simulation. In DSS enter value in any beyond the data it shows none result but agent based CBR shows near value means it is capable to solve query. Its shows with the help of diagram.

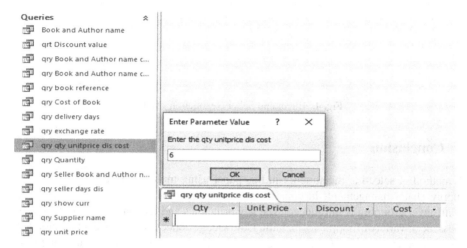

Fig. 3. Output of DSS without intelligence

According to Fig. 3 DSS is very effective but no intelligent agents are absent here. No heuristic method used for optimal solution in simple DSS .It takes a lot of time to solve complex problem. By normal DSS there are less surety for giving good result. Fault tolerance is missing in normal DSS.

Multi agent system enhance based the capability of normal DSS. It also used heuristic method for improvement of different functions. DSS respond fastly and solve complex query within period of time by using Intelligent Agents. It gives surety for better results. It also works on fault tolerance problem in normal DSS (Fig. 4).

The purchaser shall indicate the time and costs of the agreement. The provider shall convey the expense and time of effective negotiation. From Fig. 7 it concludes that the cost of time negotiating is smaller. Buyers would also be able to profit further.

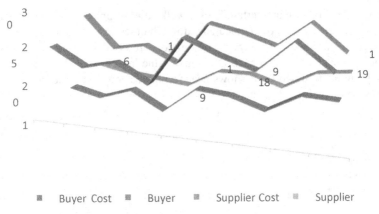

Fig. 4. Buyer and Seller Negotiation graph

7   Conclusion

A method of selecting the supplier in this paper using multi-agent technology for the supply chain management. In constructing the structure framework, the principle of multi-layer grouping is introduced. The key principles and approaches used to address the issue of variety of suppliers, including knowledge collection, assessment of intended usefulness, and the use of the time negotiating process, were discussed and analyzed. The credibility and risk dilemma dependent on the multi-agent structure is now being worked on. When you solve a problem using case-based reasoning, you retain pertinent information for future problem-solving. As part of CBR, a new problem is matched to an earlier solution, and a new case is indexed and stored according to the earlier solution so that it can be informally retrieved later. A person typically recalls a past similar problem whenever a new one arises. After doing research on several products sellers, a librarian might recall an online bookshop he purchased a few months ago. A bookseller's list provides libraries with information about the cost, discount, and delivery of a book. A reminder is sent when the price of the book they want is lower, or they get a discount or free shipping. By re-examining past experiences, humans are better able to solve problems. Let's take the previous example as an example. Cognitive psychologists agree with this statement. Analogy by learning emphasizes the mapping of a current problem description to a known problem of a domain- specific nature, while case-based strategies are used to index and match cases. AI case- based methods are based on machine-learning techniques, and CBR is one of the subfields of machine learning. In CBR, problems are solved through game playing. CBR uses a variety of methods for organizing, indexing, retrieving, revising, and utilizing past experiences. The solution obtained from the past case may be compared directly with the current case or revised if there are differences between the two.

References

1. Aamodt, A., Plaza, E.: Case-based reasoning: foundational issues, methodological variations, and system approaches. AI Commun. **7**(1), 39–59 (1994)
2. Solanki, R.: Principle of Data Mining, pp. 386–398. McGraw-Hill Publication, India (1998)
3. Abutair, H.Y., Belghith, A.: A multi-agent case-based reasoning architecture for phishing detection. Procedia Comp. Sci. **110**, 492–497 (2017)
4. Ayzenshtadt, V., Langenhan, C., Bukhari, S.S., Althoff, K.D., Petzold, F., Dengel, A.: Thinking with containers: a multi-agent retrieval approach for the case- based semantic search of architectural designs. ICAART **1**, 149–156 (2016)
5. Berman, A.F., Maltugueva, G.S., Yurin, A.Y.: Application of case-based reasoning and multi-criteria decision-making methods for material selection in petrochemistry. Proc. Inst. Mech. Eng. Part L J. Mater. Des. Appl. **232**(3), 204–212 (2018)
6. Byrski, A., Dreżewski, R., Siwik, L., Kisiel-Dorohinicki, M.: Evolutionary multi-agent systems. Knowl. Eng. Rev. **30**(2), 171–186 (2015)
7. De Mantaras, R.L., et al.: Retrieval, reuse, revision and retention in case-based reasoning. Knowl. Eng. Rev. **20**(3), 215–240 (2005)
8. Frank, E., Hall, M.A., Witten, I.H.: The WEKA Workbench. Fourth Edition (2016)
9. Haque, B.U., Belecheanu, R.A., Barson, R.J., Pawar, K.S.: Towards the application of case-based reasoning to decision-making in concurrent product development (concurrent engineering). Knowl. Based Syst. **13**(2–3), 101–112 (2000)
10. Jani, H.M., Mostafa, S.A.: Implementing case-based reasoning technique to software requirements specifications quality analysis. Int. J. Advancements Comput. Technol. **3**(1), 34–39 (2011)
11. Kwon, O., Im, G.P., Lee, K.C.: MACE-SCM: a multi-agent and case-based reasoning collaboration mechanism for supply chain management under supply and demand uncertainties. Expert Syst. Appl. **33**(3), 690–705 (2007)
12. Jiawei, H., Micheline, K., Jian, P.: Data Mining: Concepts and Techniques, 3rd Edition (2011)
13. Jagjit Singh Dhatterwal, Kuldeep Singh Kaswan, Preety"Intelligent Agent Based Case Base Reasoning Systems Build Knowledge Representation in Covid-19 Analysis of Recovery Infectious Patients" in book entitled "Application of AI in COVID 19" published in Springer series: Medical Virology: From Pathogenesis to Disease Control, July 2020, ISBN No. 978–981–15–7317–0 (e-Book), 978–981–15–7316–3 (Hard Book) https://doi.org/10.1007/978-981-15-7317-0
14. Mostafa, S.A., Ahmad, M.S., Firdaus, M.: A soft computing modeling to case-based reasoning implementation. Int. J. Comp. Appl. **47**(7), 14–21 (2012)
15. Richter, M.M., Weber, R.O.: Case-Based Reasoning. Springer-Verlag, Berlin (2016)
16. Dhatterwal, J.S., Dixit, S., Srinivasan, S.: "The role of MAS based CBRS using DM Techniques for the supplier selection. Int. J. Comp. Sci. Eng. **7**(5), 1658–1665 (2019)
17. Dhatterwal, J.S., Dixit, S., Srinivasan, S.: Implementation of a case-based reasoning system using multi-agent system technology for a buyer – seller negotiation system. Int. J. Mod. Electron. Commun. Eng. (IJMECE) **7**(3), 23–27 (2019)
18. Kaswan, K.S., Dhatterwal, J.S., Balyan, A.: Intelligent agents based integration of machine learning and case base reasoning system. In: IEEE Conference "2nd International Conference on Advance Computing and Innovative Technologies in Engineering", Galgotias University, Greater Noida (2022) https://doi.org/10.1109/ICACITE53722.2022.9823890
19. Roberta, C., Giovanni, C., Enrico, D., Andrea O.: Logic-Based Technologies for Intelligent Systems: State of the Art and Perspectives. Information (2020)
20. Thiago, P.D.H., Paulo, E.S., Anna, H.R.C., Reinaldo, A.C.B., Ramon L., de-M.: Qualitative case-based reasoning and learning. Artificial Intelligence, 283 (2020)

21. Scientific & Technology Research Volume 8, Issue 11, November 2019 ISSN: 2277- 8616
22. Koton, P.: Using experience in learning and problem solving. Massachusetts Institute of Technology, Laboratory of Computer Science, Ph.D. Thesis MIT/LCS/TR441 (1989)
23. International Joint Conference on Artificial Intelligence, Los Angeles, Calif, pp. 18–23 (1985)
24. Bain, W.M.: Case-Based Reasoning A Computer Model of Subjective Assessment Ph.D. Thesis. Yale University, Yale (1986)
25. Lebowitz: The utility of similarity based learning proceedings of the fifth national conference on Artificial intelligence, pp. 573–577 (1986)
26. Hammond, K.J.: CHEF a model of case-based planning. In: Proceeding American Association for Artificial Intelligence, AAAI-86, Philadelphia, PA, US. August (1986)
27. Acorn, Walden Acorn, T., Walden, S.: SMART: support management cultivated reasoning technology Compaq customer service In: Proceedings of AAAI -92 Cambridge. MA: AAAI Press/MIT Press (1992)
28. Kolodner, J.L.: Case Based Reasoning. Morgan Kaufmann (1993)
29. Goodman, M.: CBR in battle planning In: Proceedings of the Second Workshop on Case-Based Reasoning, Pensacola Beach, FL, US (1989)
30. Simpson, R.L.: A computer model of case-based reasoning in problem solving an investigation in the domain of dispute mediation technical report GIT-ICS-85/18, Georgia Institute of Technology, School of Information and Computer Science, Atlanta, US (1985)
31. Toorajipour, R., Sohrabpour, V., Nazarpour, A., Oghazi, P., Fischl, M.: Artificial intelligence in supply chain management: a systematic literature review. J. Bus. Res. **122**, 502–517 (2021). ISSN 0148–2963
32. Schachner, T., Keller, R., Wangenheim, F.V.: Artificial intelligence-based conversational agents for chronic conditions: systematic literature review. J. Med. Internet Res. **22**(9), e20701 (2020). PMID: 32924957; PMCID: PMC7522733 https://doi.org/10.2196/20701

Darwinian Lion Swarm Optimization-Based Extreme Learning Machine with Adaptive Weighted Smote for Heart Disease Prediction

D. Sasirega[1,2] and V. Krishnapriya[3(✉)]

[1] Sri Ramakrishna College of Arts and Science, Coimbatore, India
[2] KG College of Arts and Science, Coimbatore, Tamil Nadu, India
[3] Department of Computer Science and Cognitive Systems, Sri Ramakrishna College of Arts and Science, Coimbatore, Tamil Nadu, India
kp@srcas.ac.in

Abstract. Predicting Cardiovascular Diseases (CVD) is essential in reducing the deaths associated with different heart diseases. However, there are complexity and class imbalance problems in existing methods that degrade the overall effectiveness. Therefore, advanced frameworks are essential for heart disease detection models. This paper proposes a competent heart disease recognition framework using proficient machine learning (ML)-based methods for each clinical data analysis stage. Initially, the pre-processing stage introduces three methods for enhancing the input clinical data quality. Firstly, the Hierarchical Density-based Spatial Clustering of Applications with Noise based on the Neighbor Similarity (HDBSCAN-NS) is proposed to eliminate the data outliers without increasing the complexity. Secondly, Class Median based Missing Value Imputation (CMMVI) approach is proposed to impute the missing values based on the class distribution. Thirdly, Adaptive Weighted Synthetic Minority Oversampling Technique with Natural Neighbors (AWSMOTE-NN) is proposed to resolve the drawbacks of collinearity in the imbalanced dataset and improve the class balancing. After pre-processing, the features are extracted, and optimal feature subsets are selected using the Owl Optimization algorithm (OOA). These selected features are used by the proposed hyper-parameter optimized classifier of Darwinian Lion Swarm Optimization-based Extreme Learning Machine (DLSO-ELM) is utilized for detecting heart disease. UCI heart disease datasets are used to validate the proposed framework, and the results showed that the OOA-DLSO-ELM-based approach provides better heart disease prediction with high accuracy and low complexity.

Keywords: Cardiovascular Diseases · Adaptive Weighted Synthetic Minority Oversampling Technique · Owl Optimization algorithm · Darwinian Lion Swarm Optimization · Extreme Learning Machine

© The Author(s), under exclusive license to Springer Nature Switzerland AG 2024
S. Satheeskumaran et al. (Eds.): ICICSD 2023, CCIS 2122, pp. 13–28, 2024.
https://doi.org/10.1007/978-3-031-61298-5_2

1 Introduction

Heart disease is the most prevalent global cause of human death; over 17 million people die from heart disease, and millions are diagnosed yearly [1]. The detection of heart diseases depends on extensive labor-intensive clinical trials, which are often costly and requires more time to process the patient data. Using artificial intelligence and big data analytics, clinical decision support systems have largely employed different data mining and ML algorithms to detect and predict various diseases [2]. ML has emerged as a powerful tool in predicting and diagnosing heart disease, offering several advantages to improve prediction accuracy. These methods effectively analyze large quantities of healthcare data to identify significant features and patterns related to heart disease [3]. Hybrid models enhance predictive capabilities, leading to increased accuracy and informed decision-making in clinical settings. However, ML techniques rely on large, high-quality datasets and the limited interpretability of complex models [4]. Thus, the problems of data outliers, missing values, class imbalance problems, high dimensionality and computation complexity are common in prediction models. This paper proposes a DLSO-ELM classifier-based framework to address the challenges of existing methods, along with advanced pre-processing and feature selection methods. Initially, the pre-processing stage employs HDBSCAN-NS for outlier data removal, CMMVI for missing value imputation, and AWSMOTE-NN to solve the class imbalanced data distribution problem. To reduce the dimensionality problem, the OOA algorithm [5] selects the optimal features the proposed DLSO-ELM classifier, which is developed by optimizing the weight and bias of the ELM classifier using a hybrid optimization algorithm of DLSO. This proposed DLSO-ELM is less complex, computationally efficient and provides high accuracy for heart disease likelihood.

2 Related Works

Some of the recent studies that utilize different ML methods for CVD prediction and their limitations are analyzed. Vivekanandan et al. [6] developed a hybrid random forest with a linear model (HRFLM) for CVD prediction. Evaluated on the Cleveland Dataset, HRFLM achieved less classification error of 11.6% and high accuracy of 88.4%. Arabasadi et al. [7] proposed a highly accurate hybrid method for diagnosing coronary artery disease using genetic algorithm-optimized neural networks (GA-NN). The method achieved an accuracy of 93.85% for the Z-Alizadeh Sani dataset. Vijayashree et al. [8] proposed a PSO-SVM for diagnosing diseases in a healthcare dataset. The PSO-SVM method achieved 88.22% accuracy with the Cleveland dataset. Khourdifi et al. [9] proposed a Swarm-Artificial Neural Network (Swarm-ANN) strategy for detecting CVD to treat patients before a heart attack efficiently. It achieved a high accuracy of 95.78% for a high-dimensional healthcare dataset of CVD. Nourmohammadi-Khiarak et al. [10] used an imperialist competitive algorithm (ICA) to pick the traits that are most pertinent for diagnosing CVD. The evaluation results show an improvement in the accuracy of 94% for the Cleveland dataset. Shahid et al. [11] aim to diagnose coronary artery disease (CAD) using a highly accurate machine learning model called emotional neural networks (EmNNs) hybridized with PSO. The model achieves competitive performance with an accuracy of 88.34% on the Z-Alizadeh Sani dataset. Balasubramanian et al. [12] proposed an approach that enhances the efficiency of the adaptive neuro-fuzzy inference system (ANFIS) by incorporating a modified glow-worm swarm optimization algorithm (M-GSO) and differential evolution (DE). It achieved 98.66% accuracy for RIM-ONE dataset. Budholiya et al. [13] presented XGBoost classifier with One-Hot encoding to perform CVD detection, and Bayesian Optimization was employed to optimize the hyper-parameters. Using the Cleveland dataset, the model achieved an accuracy of 91.80%, specificity of 96.96%, sensitivity of 85.71%, and F1-score of 90.56%. Al Bataineh et al. [14] proposed a multilayer perceptron (MLP) with PSO for CVD detection. The MLP-PSO classifier yielded results with an accuracy of 84.6%, and AUC (Area under the Curve) of 84.8% using the Cleveland dataset. El-Shafiey et al. [15] studied the prediction of CVD using a hybrid genetic algorithm (GA) and PSO-optimized approach called GAPSO-RF. It achieved accuracy of 95.60% for Cleveland dataset and 91.40% for the Statlog dataset. The limitations of machine learning methods for predicting CVD include the generalizability of the proposed models to different datasets and their reliance on specific algorithms and features. To overcome these constraints, the proposed method utilizes DLSO-ELM in CVD prediction.

3 Methods

The suggested method for predicting heart disease includes three important processes: pre-processing, feature learning and classification. The overall work method of this suggested approach is shown in Fig. 1.

3.1 Data Pre-processing

In this pre-processing step, eliminating the outliers, imputing missing values and balancing the class are vital tasks to improve the data quality. The three methods of HDBSCAN-NS, CMMVI and AWSMOTE-NN are utilized for pre-processing.

Outliers Removal Using HDBSCAN-NS. In this proposed HDBSCAN-NS method, the neighbor similarity is measured based on Cover Tree and triangle inequality to retrieve neighbors and avoid unnecessary distance computations. In the HDBSCAN-NS method, a single input parameter *MinPts* denoting the density threshold or minimum object points is assigned a value for smoothing the density estimates. The resulting density-based cluster hierarchy's densities variations will then correlate to multiple radius ranges ε. For formulating the density-based hierarchy correctly based on the value of *MinPts*, the DBSCAN notions are defined to suit the HDBSCAN. The core distance of an object $x_i \in X$ is defined as $d_{core}(x_i)$ is based on the *MinPts* and denotes the distance from x_i to its nearest neighbor *MinPts*. An object $x_i \in X$ is defined as an ε-core object for each value of ε which are larger than or equal to the core distance of x_i with respect to *MinPts*, i.e., if $d_{core}(x_i) \leq \varepsilon$. .. The mutual reachability distance (MRD) between two objects x_i and x_j in X with respect to *MinPts* can be defined as $d_{mreach}(x_i, x_j) = \max\{d_{core}(x_i), d_{core}(x_j)d(x_i, x_j)\}$. The MR graph, G_{MinPts}, is a full graph in which each edge's weight represents the MRD between the specified pair of items. The vertices of the MR graph are the objects of X. As a result, a cluster C becomes a non-empty maximum subset of X such that every pair of objects in C is related by density with respect to *MinPts*.

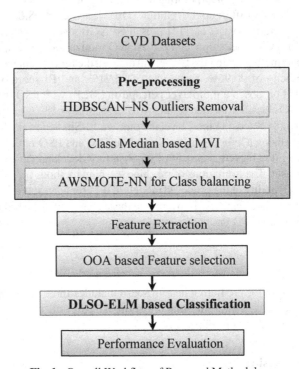

Fig. 1. Overall Workflow of Proposed Methodology.

Let $G_{MinPts,\varepsilon} \subseteq G_{MinPts}$ be the graph achieved by eliminating all the edges from G_{MinPts} having weights larger than ε. Based on the above definitions, it is conveyed that the clusters based on ε and *MinPts* are associated modules of ε-core objects in $G_{MinPts,\varepsilon}$ while the lasting items are noise. Subsequently, the dividers for $\varepsilon \in [0, \infty]$ are formed in a nested, tiered manner by eliminating the edges in declining weights from G_{MinPts}. When X denotes a set of n objects labeled in a metric space by $n \times n$ pairwise estimations. The grouping of this data attained by the DBSCAN process based on ε and *MinPts* will be identical to the clusters achieved by Single-Linkage over the transformed space of MRD, which means dividing the subsequent dendrograms at level ε of its measure and handling all the ensuing singletons with $d_{core}(x_i) > \varepsilon$ as noise. The object o is labeled as noise below the lower bound l, which corresponds to the object's core distance, and must be represented by the density-based cluster hierarchy. The HDBSCAN-NS determines the neighborhood of each data point and computes the neighbor similarity. It is computed based on the tiered wrapper tree for the source dataset. To represent the unlabeled or unprocessed data, the unprocessed data is initialized, the neighbors are set for each point to NULL, and the labels are set for each point to -2. The suggested algorithm creates the query tree TQ by choosing the seeds num data point or 1000 or more, from the unprocessed data. Following a concurrent retrieval of each query point's closest neighbors using the cover tree, the HDBSCAN-NS characteristics are utilized to identify the portion of outliers. For a point $p \in P$, if the ε-neighborhood of the second point $|N_{2\varepsilon}(p)| \geq MinPts$, then $\forall q \in N_{2\varepsilon}(p)$, q denotes the non-core-point. Then the core points p will be combined into further core samples to form different clusters. The border points are assigned to the nearest clusters, while the out-of-bound points will be assigned to the outlier cluster, which will be removed from the input data.

Class Median-Based Missing Value Imputation. CMMVI method is based on calculating the class median for each class, and then using the gaps between it and the other observable data to establish a threshold for the subsequent imputation. It consists of two modules; the first establishes the imputation threshold based on the separations between the class median and the relevant data samples, and the second applies the MVI threshold. The first module's goal is to determine the missing value imputation threshold for the following module and includes the following steps.

Step 1: For an partial dataset D that includes magnitudes X and classes Y.

Step 2: D_i, represents the i-th class (i = 1 to X) of D, and is classified into whole $(D_{i_{complete}})$ and partial subgroups $(D_{i incomplete})$, where partial subsets contain absent values.

Step 3: The class center, $cent(D_i)$, and standard deviation, std_i is calculated for the i-th class of $(D_{i_{complete}})$ then the Euclidean distance between $cent(D_i)$, and each data points is calculated in class i.

Step 4: The median distance is used as the threshold (Ti) for that specific class by taking into account the distances between the class center and individual data samples within class i. Steps 3 and 4 are then repeated up until the threshold for each class is established.

The imputation of single and multiple missing values for Module 2 contains 3 stages.

Step1: The incomplete dataset $(D_{i incomplete})$, in the case of the i-th class, which consists of *Num_miss* data samples, based on the number of missing values present. If there is only one absent value, proceed to Step 2. However, if multiple missing values exist, move on to Step 3.

Step 2: When the data j (j = 1 to *Num_miss*) contains for $D_{i_{complete}}$ only 1 missing value of j.

Step 3: The feature values related of $cent(D_i)$ are imputed for these missing values for numerous absent data j. For threshold T_i, comparison, the distance between the imputed data j and $cent(D_i)$ is estimated.

Class Imbalance Problem Using AWSMOTE-NN. In AWSMOTE-NN model, an adaptive weight is used in adaptable space to handle the drawbacks of collinearity in the imbalanced dataset. To improve the distribution and reduce collinearity, the approach introduces variable weights obtained from the estimate vector υ. Variables represent the influences of variable weights in a two-dimensional space X_1 and X_2). The weights w_1 and w_2 correspond to the weights of variable 1 and variable 2, correspondingly. When generating an artificial sample $(\tilde{x})$ using SMOTE, the line segment between the minority sample (x) and its adjacent sample (x_p) is considered. The generated samples shall no longer be simple linear groupings of x and x_p p due to the introduction of variable weights. Depending on the values and directions of the weights, the generated samples deviate from linear combinations. When $w_2 > w_1 > 0$, it suggests that the weight

w_2 has a significant impact on the sample creation. In this case, the generated sample $(\tilde{x}_1)$ is closer to the nearest neighbor x_p. Similarly, when $w_1 > w_2 > 0$, the weight w_1 has a greater influence, and the created sample $(\tilde{x}_2)$ is more persuaded towards w_1. Contrariwise, when both w_1 and w_2 are less than 0, the course of the created sample $(\tilde{x}_3)$ is opposite to $(\tilde{x}_1)$.

The element of the j-th variable in $(\tilde{x}_{i,j})$.for $(\tilde{x}_i)$ the newly generated sample is represented as;

$$\tilde{x_{ij}} = x_{i,j} + n \times (x_i - x_i^k) \times w_j \tag{1}$$

The random number ranging from [0,1] is denoted as n and one of K neighbors of the minority sample x_i is represented as x_i^k. The weight of the variable where $j \in [1, p]$ is denoted as w_j. The effect of each variable weight on the selection of samples is equal when $w_1 = \cdots = w_p = w$. On the line segments connecting the neighbor and the minority instance, x_i, a new sample is produced x_i^k based on the neighbour. Given that the weights of the variables are not equal, the sample generation for the influence of the variable is different. For instance, if the variable $w_1 > w_2 \ldots, w_p$ variable weights, then it is reasonable to assume $\tilde{x}_{i1}$ shows a significant impact in $\tilde{x}_{ij}$, thus the generation of $\tilde{x}_i$ is affected by $\tilde{x}_{i1}$. To some extent effectively optimize the distribution of the minority class, the variable weight is introduced. When incorporating variable weights, creating a new sample no longer adheres to the linear path between a minority sample and its closest neighbor. During the oversampling of boundary samples, utilizing SMOTE can amplify the overlapping areas between classes, especially when the neighbouring point resides on the classification boundary. Incorporating variable weights, however, makes the generated sample less likely to be oriented toward nearby samples, somewhat reducing the degree of class overlapping.

The SMOTE method is improved based on the weights of variables. For each minority class sample x_i with p variables, the natural neighbors are applied for outlier detection. The Euclidean distance (Dis_i) is used to calculate the distance between the two instances

$$Dis_i = x_i - NN = \sqrt{\sum_{j=1}^{p} \left(x_{ij} - NN\right)^2} \tag{2}$$

A new instance is generated for each of the elements of the jth variable is defined as

$$\widetilde{NNs} = x_{ij} + w_j \times n_i \times Dis_i \tag{3}$$

where $n_i \in [0, 1]$ is a arbitrary integer, w_j is the weight of each variable and $j = 1,2,\ldots p$.

The AWSMOTE is used for calculating Case Weight involving calculating variable weights and minority sample weights of natural neighbors. The number of new samples generated for each of the natural neighbors (NNs) in the minority class data μ is determined based on their weights, and the method is used to generate these samples in steps 4–12. The algorithm searches the minority class data μ. NNs are the data points similar or close to each other regarding their attribute values in the minority class. The outliers are removed from the minority class data to avoid distorting the data distribution and prevent the generation of additional outliers that to maintain the data distribution integrity. The minority class data μ is randomized to prevent any bias introduced by order of the data. For each base sample (selected from μ), a certain number of synthetic samples (SN) are generated. One of the NNs is selected for each base sample, calculates the difference between the attributes of the base sample and the selected neighbor for each attribute dimension, and is scaled by a random number between 0 and 1. Then the scaled differences are added to the attribute values of the base sample, resulting in the generation of new synthetic samples.

3.2 Feature Selection Using OOA

This study uses the wrapper method of Owl Optimization Algorithm (OOA) for the feature section by optimally ranking the feature subsets. This OOA is based on the decoy behavior exhibited by burrowing owls in the presence of predators or other threats near their nests [8]. The algorithm contains 5 parameters $(PP_n, SP_n, DP_{nest}, R_1, R_2)$ to tune the first 3 values are cluster process and nest renewal parameters, and the other two are generated random values with uniform distribution in the range [0, 1].

Stage 1: Initialization: The existing problem, such as the optimization problem, decision parameters, is defined. Further, the OOA parameters are adjustable such as O_n represents the number of Owls, PP_n denotes the number of primary branches, SP_n is the number of secondary perches, $\max\limits_{iter}$ is the highest iteration number and DP_{nest} is the deprecated nests' percentile. It is crucial to observe that the number of primary branches and subordinate branches should match the number of owls, as stated $O_n = PP_n + SP_n PP_n$.

Stage 2: Initializing Owl's Position: Regarding the boundary conditions of a D-dimensional space, the owls are positioned randomly and represented as a matrix;

$$O = \begin{matrix} O_{1,1} & \cdots & O_{1,D} \\ \vdots & \ddots & \vdots \\ O_{O_n,1} & \cdots & O_{O_n,D} \end{matrix} \qquad (4)$$

where the individual in the O_n position with dimensions D is represented as $O_{O_n,D}$.

Stage 3: Sorting: The vector is generated by obtaining the fitness value for the problem for each owl;

$$F_{ord} = \begin{bmatrix} F_{ord1} \\ \vdots \\ F_{ord2} \end{bmatrix} \qquad (5)$$

Subsequently, a sorting algorithm is employed to arrange and form clusters according to the attained fitness values, referred to as F_{ord}^{sort}. Each cluster is assigned specific primary and secondary perches. Furthermore, random numbers $R_1 and R_2$ are generated for this stage.

Stage 4: Position Update: According to the decoy behavior of owls, new positions are generated as O_{new} in search space. Initially, it moves based on two distinct positions within the cluster: the subsequent better perch and the optimal position within the cluster. If the new position is not better than its previous one and takes position in the best perch or if that is not found better, it returns and remains in its current position. According to the deprecated nests' percentile, the primary branches are also updated as a new perch O_{new}^{sort} obtained from the better perch in position search. If the new position is not preferable to the prior one, the perch is allocated at its existing location.

Stage 5: Termination: The process terminates when the maximum number of iterations is attained by repeating the location search from the third and fourth stages. Following all assessments, a fresh sorting is carried out for the following generation.

The multi-objective variant was designed by integrating concepts like dominance, non-dominance, Pareto Set, and Pareto Front (PF). Due to the sorting process, it is worth highlighting that sorting is employed based on a specific objective function.

3.3 CVD Prediction Using DLSO-ELM

In this proposed DLSO-ELM method, the LSO algorithm is combined with the Darwinian Theory to develop the DLSO algorithm, which is used to optimally select the weight and bias values of the ELM classifier. The LSO algorithm draws inspiration from lions' social behavior, which involves cooperation and antagonism within their groups [16]. To regulate the speed at which the LSO algorithm converges, an adult lion proportion factor known as Ψ is introduced. Ψ is a positive random number chosen from 0 to 1. To ensure a faster convergence, it is specified that Ψ should be set to a value below 0.5. Once an approximate position of the optimal solution is identified, it becomes essential to strengthen the optimizer's capability for local exploration by introducing a disruption factor,κ_f, which enables the lionesses to engage in broader exploration. As the optimization process advances, the lionesses' activity range gradually shrinks, allowing the algorithm to shift towards a localized search in the proximity of the optimal solution that enhances the accuracy. A balance is achieved between global and local search by incorporating the disruption factor κ_f, leading to enhanced convergence speed and prevention of premature convergence. The specific expression for the disruption factor κ_f, is defined as

$$\kappa_f = L_{a1}.\exp\left[-30.\frac{g}{Max_{iter}}\right] \qquad (6)$$

where the maximum iteration is denoted as Max_{iter}, the g-th current iteration as g, and the step value of the lionesses activity range represented as L_{a1}, by

$$L_{a1} = r_1.(\overline{a_{max}} - \overline{a_{min}}) \qquad (7)$$

where a random number in the range [0,1] is denoted by r_1 and identified as the lioness' step's controlling element. The smallest and largest mean values for each dimension are represented by ($\overline{a_{max}}$ and $\overline{a_{min}}$), respectively. The interaction between lion cubs and the lion king for food and their learning process from lionesses during hunting is constrained within a specific range. To enable a wider exploration in the initial stages and a subsequent focused search, a disruption factor, φ_m is introduced that gradually reduces the activity range of lion cubs, enabling them to initially search for prey over a broader area and gradually shift towards a more localized search around the prey. The disruption factor φ_m is expressed as

$$\varphi_m = L_{a2} \cdot \frac{Max_{iter} - g}{Max_{iter}} \tag{8}$$

where the step value in the array of lion cubs' motion is represented by L_{a2} and defined as

$$L_{a2} = r_2 \cdot (\overline{a_{max}} - \overline{a_{min}}) \tag{9}$$

where a random number in the array [0,1] is denoted by r_2 and recognized as the control factor of the lion cub's step. Each lion position in the LSO algorithm indicates a possible solution to the problem under consideration, and the calibre of the prey reflects the calibre (fitness) of the corresponding solution. An initial population P called pride containing δ solutions (lion locations) will be arbitrarily produced in the search space for a D-dimension GOP $\prod_h^D (\overline{a_{max,j}} - \overline{a_{min,j}})$ in the beginning. The following matrix can represent all lion positions.

$$a = \begin{matrix} a_{1,1} & a_{1,2} & L & L & a_{1,D} \\ a_{2,1} & a_{2,2} & L & L & a_{2,D} \\ M & M & M & M & M \\ M & M & M & M & M \\ a_{n,1} & a_{n,2} & L & L & a_{n,D} \end{matrix} \tag{10}$$

From Eq. (10) L represents the leader, M represents the members and $a_{i,j}$ denotes ith lion in jth dimension, for each D-dimensional vector $(a_i = a_{i,1}, a_{i,2}, K, a_{i,D})$, where K = 1,2,...n denotes the number of lions that represents the position of ith lion generated by

$$a_{i,j} = a_{min,j} + rand(0, 1) \cdot (a_{max,j} - a_{min,j}) \tag{11}$$

where j = 1, 2, L, n and i = 1, 2, L, n. The homogeneously distributed arbitrary integer in range (0, 1) is represented by rand (0, 1). The maximum and minimum bounds of the jth dimension are denoted as $a_{max,j}$ and $a_{min,j}$, respectively. Thus the initialization process ends. Hence, the quantity of lion king and lioness, $nleader(2 \le nleader \le \frac{n}{2})$, i.e., the adult lions are given by

$$nleader = n, \psi \tag{12}$$

The $n - nleader$ represents the total of lion cubs.

For each lion position, the fitness value is calculated by assigning user-defined fitness functions are used to translate decision variable values (solution vector) into equivalent fitness values are given by;

$$Ff = \begin{bmatrix} Ff_1([a_{1,1}, a_{1,2}, L, a_{1,D}]) \\ Ff_2([a_{2,1}, a_{2,2}, L, a_{2,D}]) \\ M \\ M \\ Ff_n([a_{n,1}, a_{n,2}, L, a_{n,D}]) \end{bmatrix} \tag{13}$$

The fitness value associated with the position of each lion reflects the quality of the prey they have searched for and can be categorized as optimal prey, which is controlled by the lion king, standard prey, which is obsessed by the lionesses; and tiny prey, which the lion cubs possess. The fitness value also indicates the probability of survival for each lion. The lion king ensures priority access to prey over other lions, and it may move within the sort of the best food, represented by the position with the lowest fitness. The new location of the lion king is given by

$$a_i(g + 1) = wbest(g).(1 + \beta.\|lbest_i(g) - wbest(g)\|) \tag{14}$$

where the β denotes the normally distributed arbitrary integer [0,1], g is the current iteration, $wbest(g)$ is the globally best location of the pride at the current iteration and $lbest_i(g)$ is the historically best location of the i-th lion at the present iteration.

The typical hunting behavior of lionesses involves identifying the position of prey, surrounding them, and then launching an attack. During this hunting behavior, lionesses often cooperate with another selected lioness from their group. In this scenario, the new positions of the lionesses are determined as

$$a_i(g + 1) = \frac{lbest_i(g) + lbest_c(g)}{2}.\left(1 + \kappa_f.\beta\right) \tag{15}$$

where the collaboration lioness is currently positioned in its historically optimal position when represented as $lbest_c(g)$ and the moving range distribution factor by κ_f.

Three different circumstances may occur while lion cubs are engaging in dynamic hunting. The lioness that the lion cubs follow is picked at random from the group of lionesses, therefore the exact coupling between the lioness and the lion cubs is determined at random. The new positions of the lion cubs are determined by

$$a_i(g + 1) = \begin{cases} \frac{wbest(g) + lbest_i(g)}{2}.(1 + \varphi_m.\beta), u \leq \frac{1}{3} \\ \frac{lbest_v(g) + lbest_i(g)}{2}.(1 + \varphi_m.\beta), \frac{1}{3} < u < \frac{2}{3} \\ \frac{\overline{wbest(g)} + lbest_i(g)}{2}.(1 + \varphi_m.\beta), \frac{2}{3} \leq u \leq 1 \end{cases} \tag{16}$$

where $\overline{wbest(g)}$ is the place where the i-th lion cub gets ejected from the hunting territory, the moving range distribution factor is denoted as φ_m, and the present iteration's optimal place for lion cubs historically is indicated as $lbest_v(g)$. The arbitrary integer in the range [0,1] is denoted as u. The search space for nomadic lion cubs is expanded, improving global exploration capacity. $\overline{wbest(g)}$ is defined as;

$$\overline{wbest(g)} = \overline{a_{max}} + \overline{a_{min}} - wbest(g) \tag{17}$$

The new swarm and lion cub production and deletion processes are now included in the application of the Darwinian theory of survival. The developing process, in which the current swarms are allowed to create a new swarm with a constant probability, is carried out at each algorithmic step. Following generation, selection/deletion is started to gather the moving swarms and eliminate the stationary swarms. By assessing the new swarm's fitness and determining the ideal placements, the evolution process is finished. After updating the positions, the swarm only creates a new lion cub when the best fitness is attained within the predetermined iterations and deletes the old poor fitness lion cub.

Building Fresh Swarms and Lions. A new swarm can be created with a predetermined minimum population when the initial iterations produce the condition that no lion cub is killed in a swarm (i.e. $N_{kill} = 0$) and the number of swarms does not reach the maximum limit. Two existing swarms could be evolved into a new swarm through iterations, up until the maximum number of swarms is reached. When the maximum number of swarms is reached, the only way to create new swarms is to eliminate an existing swarm, i.e., it must have $N_{kill} = 0$ to continuing to create fresh swarms. In addition to the number of swarms N_s, the newly formed swarms will have probability $p(f/N_s)$ through uniform arbitrary integer $f \in [0, 1]$. The freshly created swarm has all of the characteristics of two-parent swarms in equal measure, and the single parent's unique dominating characteristics can be chosen to enhance the DLSO design. The procedure again changes to creating another new swarm once the new swarm has achieved the global best fitness.

Removing Swarm. Swarms are only removed when their overall number of lion cubs falls below a set threshold. The lion cub population of the swarm is constrained so that $X_{min} \leq X \leq X_{max}$. So, when $X < X_{min}$, the swarm is eliminated from the solution set.

Removing a Lion Cub. A search monitor is constructed with predetermined minimum and maximum values for the removal of a lion cub. The search counter, SC, starts at zero when a new swarm is created and adds 1 anytime the fitness value of the lion cubs doesn't increase. The lion cub that consistently produces the worst fitness, even after numerous evolutions, is destroyed, and SC is used to track the number of evolutions that produced unfavourable fitness. The highest critical threshold for this timer is fixed at Threshold SC_c^{max}. In order to shorten the amount of time needed to track fitness improvement, the lion cub that surpasses SC_c^{max} is eliminated from the pack and the search monitor is reset to a value that is closer to SC_c^{max} or zero. The swarms' fitness will increase as a result of this reduction without the delay tolerance suffering. Depending on the value of N_{kill}, which is as follows, this reset value for the search monitor is chosen.

$$SC_c(N_{kill}) = SC_c^{max}\left[1 - \frac{1}{N_{kill} + 1}\right] \tag{18}$$

Therefore, the proposed DLSO-ELM algorithm aims to achieve high accuracy with minimal hidden layer nodes in the ELM prediction model. The ELM model optimally produces the hidden layer's input nodes, weight and bias vector. The number of hidden layer neurons is represented as *HL*, input layer neurons as *IL*, and output layer neurons as *OL*. The weight matrix is represented as ξ and the bias vector as θ. The weight matrix of hidden layer neurons is given by

$$\xi = \begin{matrix} \varepsilon_{11} & \varepsilon_{12} & \cdots & \varepsilon_{1n} \\ \varepsilon_{21} & \varepsilon_{22} & \cdots & \varepsilon_{2n} \\ \vdots & \vdots & \ddots & \vdots \\ \varepsilon_{HL1} & \varepsilon_{HL2} & \cdots & \varepsilon_{HLn} \end{matrix} \tag{19}$$

The bias vector of hidden layer neurons θ and given by

$$\theta = \begin{bmatrix} b_1 \\ b_2 \\ \vdots \\ b_{HL} \end{bmatrix}_{HL \times 1} \tag{20}$$

It is important to guarantee that the matrix ξ and θ are resolved by Eq. (21) for the optimization problem.

$$G = G(\xi, \theta) = [a_{ij}]_{N \times HL} = \begin{bmatrix} f(w_1.a_1 + b_1) & \cdots & f(w_{HL}.a_1 + b_{HL}) \\ \vdots & \ddots & \vdots \\ f(w_1.a_N + b_1) & \cdots & f(w_{HL}.a_N + b_{HL}) \end{bmatrix} \tag{21}$$

where f(.) is the activation function. The solution for the optimization problem is given by

$$\widehat{OL} = \underset{OL}{\arg\min} \|GOL - T\| \tag{22}$$

$$\widehat{OL} = G^\dagger T \tag{23}$$

$G^\dagger = (G^T G)^{-1}$ is the hidden layer output matrix G's Moore-Penrose generalized inverse matrix. When ($G^T G$ is not singular, $G^\dagger = (G^T G)^{-1} G^T$. Based on Eq. (22), a ridge regression contains a constraint factor $\Im$. Then $\widehat{OL}$ is calculated as

$$\widehat{OL} = \frac{1}{\Im} + \left(G^T H\right)^{-1} G^T T \tag{24}$$

The main steps of the DLSO-ELM algorithm are as follows;
Step 1: Pre-process the original dataset using OOA.
Step 2: Set the objective function of DLSO as accuracy between predicted and actual values. Define the (Max_{iter}), the position of lions (a_i), the adult lion's proportion factor (Ψ), and the search range for the hidden layer nodes [Nod_{min}, Nod_{max}] Nod_{min} to the optimal number of nodes (Nod_{opt}) and Nod_c to the current number of nodes. Randomly

generate the initial weight matrix ξ and bias vector θ. Calculate the accuracy and set it as the optimal result ω.

Step 3: Calculate the size of the location vector a_i for the ith lion using the equation $(a_i + 1) \times Nod_c$ and the number of input layer nodes, indicated as "*IL.*" Add the random numbers r_1 and r_2 to the elements of the initial position vector a_i.

Step 4: Update the positions of the lion king, lionesses, and cubs using Eq. (16), respectively, based on their current positions.

Step 5: After performing step 4 five times, adjust the lion-ruler's position to its most advantageous setting.

Step 6: Verify that the termination condition has been satisfied. The goal function value is less than a predetermined threshold, which is the termination condition. Step 7 should be taken if the prerequisite is met. If not, return to step 5.

Step 7: Divide the lion-ruler's position vector into the bias vector θ and weight matrix ξ should be used to evaluate the present fitness function value ω. Update, and (Nod_{opt}) with the current values if the objective function value is larger than it was previously. Next, move on to step 8. Otherwise, go straight to step 8.

Step 8: Increment (Nod_c) by 1. If Nod_c is greater than Nod_{max}, output (Nod_{opt}), ξ, and θ as the optimal hidden layer nodes, weight matrix, and bias vector, respectively. Otherwise, return to step 3. Thus, the proposed DLSO-ELM-based heart disease prediction procedure is cost-effective in improving the prediction accuracy.

4 Performance Evaluation

The implementations are performed using MATLAB R2021 in an Intel i5 system with Windows 10 OS. The proposed classification-based OOA-DLSO-ELM heart disease prediction approach is implemented and evaluated for UCI datasets namely Cleveland, Hungary, Switzerland and VA Long Beach. Table 1 shows the performance comparison of the proposed OOA-DLSO-ELM with existing classifiers.

Table 1. Performance of proposed OOA-DLSO-ELM with existing classifiers.

Methods/ Metrics	ELM	LSO-ELM	DLSO-ELM	Proposed OOA-DLSO-ELM
Cleveland Dataset				
Accuracy (%)	82.4	85.6	88.7	99.2806
MCC	0.88	0.91	0.89	0.9864
Precision (%)	80.4	81.6	79.5	99.6262
Recall (%)	78.3	85.2	86	99.2000
F-measure (%)	83.9	89.7	92	99.4126
Time (seconds)	10.6	9.9	7.6	4.2842

(continued)

Table 1. (*continued*)

Methods/ Metrics	ELM	LSO-ELM	DLSO-ELM	Proposed OOA-DLSO-ELM
Hungarian Dataset				
Accuracy (%)	89.5	90.4	93.5	99.4536
MCC	0.79	0.88	0.92	0.9867
Precision (%)	85.9	88.4	90.2	99.0521
Recall (%)	90.4	92	93.5	99.6169
F-measure (%)	87.2	90.8	92.4	99.3337
Time (seconds)	11.9	10.2	9.4	5.0325
Switzerland dataset				
Accuracy (%)	78	79.6	83	98.7342
MCC	0.88	0.91	0.90	0.9884
Precision (%)	90.3	91.4	92.1	99.3569
Recall (%)	91.6	90.4	91.8	99.0476
F-measure (%)	88.6	85.5	89	99.2020
Time (seconds)	15.4	12.8	11.7	4.0208
VA long beach dataset				
Accuracy (%)	77.6	80.3	83.7	99.0909
MCC	0.76	0.82	0.85	0.9816
Precision (%)	92.9	91.1	93.5	99.5272
Recall (%)	90.8	91.4	93.1	99.2481
F-measure (%)	87.6	88	89.5	99.3875
Time (seconds)	15	14.8	13	5.4029

Table 1 shows that the proposed method, OOA-DLSO-ELM, outperforms ELM, LSO-ELM, and DLSO-ELM in terms of accuracy, MCC, precision, recall, and F-measure and demonstrates improved efficiency by significantly reducing the processing time. These results highlight the effectiveness and superiority of the proposed method in classification tasks, showcasing its potential for various real-world applications. Table 2 shows the evaluation of the suggested OOA-DLSO-ELM method with existing procedures on the Cleveland dataset. The comparison results in Table 2 indicate that the proposed OOA-DLSO-ELM-based disease classification model has achieved good performance than the models in the literature. These results highlight the dominance of the suggested OOA-DLSO-ELM method in achieving higher accuracy and better performance than the evaluated methods.

Table 2. Performance comparison of OOA-DLSO-ELM with existing techniques.

Methods/ Metrics	Accuracy (%)	MCC	Precision (%)	Recall (%)	F-measure (%)	Time (s)
SVM	94.5	0.76	87.2	72.6	87.3	12.57
Naïve Bayes	93.6	0.84	85.4	74.9	80.6	25.74
Random Forest	94	0.89	88	76	81.5	11.6
Decision Tree	95.7	0.79	82.7	75.3	79.4	12.4
HRFLM [6]	88.4	0.81	90.1	71.4	90	13.8
GA-NN [7]	93.85	0.88	89	75.6	75.4	11.76
PSO-SVM [8]	88.22	0.91	86.4	77	74.8	35.9
PM-LU-NN [9]	95	0.86	80.3	81	80.6	28.7
PSO-ANFIS [10]	88.34	0.89	92.3	80.6	84	11.12
FCM-DBN-TBSA [11]	93.4	0.92	85.6	79.3	88.7	6.3
DE-GSO-ANFIS [12]	98.66	0.86	83.9	78.5	85.7	24.56
XGBoost-BO-OH [13]	91.8	0.76	85	79.3	90.5	48.7
MLP-PSO [14]	80.8	0.69	81.9	88.3	84.4	25.8
GAPSO-RF [15]	95.6	0.84	97.4	92.6	91.3	25.59
Proposed OOA-DLSO-ELM	**99.2806**	**0.986**	**99.6262**	**99.200**	**99.4126**	**4.2842**

5 Conclusion

This manuscript offered an effective disease classification model using efficient pre-processing and advanced ML classifier methods. The proposed framework utilized OOA-based feature selection over the pre-processed information to reduce the dimensionality and propagate the top attributes to the classifier. Then the proposed ML classifier of DLSO-ELM is used for individually categorizing the patient data. Evaluated over the benchmark instances, the proposed OOA-DLSO-ELM-based model improved the disease classification accuracy and reduced the model's complexity. In the future, the possibility of including multi-source clinical datasets and hybrid feature selection methods will be investigated to improve the disease detection rate.

References

1. Tsao, C.W., Aday, A.W., Almarzooq, Z.I., Anderson, C.A., Arora, P., Avery, C.L.: Heart disease and stroke statistics—2023 update: a report from the American Heart Association. Circulation **147**(8), e93–e621 (2023). https://doi.org/10.1161/CIR.0000000000001123
2. Ahsan, M.M., Siddique, Z.: Machine learning-based heart disease diagnosis: a systematic literature review. Artif. Intell. Med. **102289** (2022). https://doi.org/10.1016/j.artmed.2022.102289
3. Petch, J., Di, S., Nelson, W.: Opening the black box: the promise and limitations of explainable machine learning in cardiology. Can. J. Cardiol. **38**(2), 204–213 (2022). https://doi.org/10.1016/j.cjca.2021.09.004

4. Quer, G., Arnaout, R., Henne, M., Arnaout, R.: Machine learning and the future of cardiovascular care: JACC state-of-the-art review. J. Am. Coll. Cardiol. **77**(3), 300–313 (2021). https://doi.org/10.1016/j.jacc.2020.11.030

5. de Vasconcelos Segundo, E.H., Mariani, V.C., dos Santos Coelho, L.: Metaheuristic inspired on owls behavior applied to heat exchangers design. Therm. Sci. Eng. Prog. **14**, 100431 (2019). https://doi.org/10.1016/j.tsep.2019.100431

6. Vivekanandan, T., Iyengar, N.C.S.N.: Optimal feature selection using a modified differential evolution algorithm and its effectiveness for prediction of heart disease. Comput. Biol. Med. **90**, 125–136 (2017). https://doi.org/10.1016/j.compbiomed.2017.09.011

7. Arabasadi, Z., Alizadehsani, R., Roshanzamir, M., Moosaei, H., Yarifard, A.A.: Computer aided decision making for heart disease detection using hybrid neural network-genetic algorithm. Comput. Methods Programs Biomed. **141**, 19–26 (2017). https://doi.org/10.1016/j.cmpb.2017.01.004

8. Vijayashree, J., Sultana, H.P.: A machine learning framework for feature selection in heart disease classification using improved particle swarm optimization with support vector machine classifier. Program. Comput. Softw. **44**, 388–397 (2018). https://doi.org/10.1134/S0361768818060129

9. Khourdifi, Y., Baha, M.: Heart disease prediction and classification using machine learning algorithms optimized by particle swarm optimization and ant colony optimization. Int. J. Intell. Eng. Syst. **12**(1), 242–252 (2019). https://doi.org/10.22266/ijies2019.0228.24

10. Nourmohammadi-Khiarak, J., Feizi-Derakhshi, M.R., Behrouzi, K., Mazaheri, S., Zamani-Harghalani, Y., Tayebi, R.M.: New hybrid method for heart disease diagnosis utilizing optimization algorithm in feature selection. Heal. Technol. **10**, 667–678 (2020). https://doi.org/10.1007/s12553-019-00396-3

11. Shahid, A.H., Singh, M.P.: A novel approach for coronary artery disease diagnosis using hybrid particle swarm optimization based emotional neural network. Biocybern. Biomed. Eng. **40**(4), 1568–1585 (2020). https://doi.org/10.1016/j.bbe.2020.09.005

12. Balasubramanian, K., Ananthamoorthy, N.P.: Improved adaptive neuro-fuzzy inference system based on modified glowworm swarm and differential evolution optimization algorithm for medical diagnosis. Neural Comput. Appl. **33**, 7649–7660 (2021). https://doi.org/10.1007/s00521-020-05507-0

13. Budholiya, K., Shrivastava, S.K., Sharma, V.: An optimized XGBoost based diagnostic system for effective prediction of heart disease. J. King Saud Univ. Comp. Inform. Sci. **34**(7), 4514–4523 (2022). https://doi.org/10.1016/j.jksuci.2020.10.013

14. Al Bataineh, A., Manacek, S.: MLP-PSO hybrid algorithm for heart disease prediction. J. Personalized Med. **12**(8), 1208 (2022). https://doi.org/10.3390/jpm12081208

15. El-Shafiey, M.G., Hagag, A., El-Dahshan, E.S.A., Ismail, M.A.: A hybrid GA and PSO optimized approach for heart-disease prediction based on random forest. Multimedia Tools Appl. **81**(13), 18155–18179 (2022). https://doi.org/10.1007/s11042-022-12425-x

16. Liu, S.J., Yang, Y., Zhou, Y.Q.: A swarm intelligence algorithm — lion swarm optimization. Pattern Recogn. Artif. Intell. **31**(5), 431–441 (2018). https://doi.org/10.16451/j.cnki.issn1003-6059.201805005

Alzheimer's Disease Detection Using Convolution Neural Networks

M. Swapna, M. Ravali, G. Pavani, M Shiva Durga Prasad, V Pradeep Kumar[✉] [iD],
and Ashok Kumar Nanda

Department of Computer Science and Engineering, B V Raju Institute of Technology, Narsapur,
Medak, Telangana, India
{19211A05E9,19211a05f4,20215a0515,19211a05h2,
pradeepkumar.v}@bvrit.ac.in

Abstract. Millions of people throughout the world are afflicted by the progressive neurological condition known as Alzheimer's Disease (AD). Analyzing MRI brain pictures is one of the most promising methods for identifying AD. Deep learning methods have become effective tools for diagnosing AD from MRI scans in recent years. An overview of deep learning-based methods for AD identification using MRI scans is given in this research, along with an explanation of the various deep learning architectures, including convolutional neural networks (CNNs) and auto-encoders. We explore the difficulties in detecting AD using deep learning, such as the necessity for big and diverse datasets and the possibility of bias and overfitting. MRI images have been used in recent studies to identify AD using deep learning, including the utilization of adversarial training and transfer learning. The proposed work will give a good accuracy where training accuracy is 86.34% and validation accuracy is 86.45% on the test data with very small misclassifications on normal and very mild demented.

Keywords: Alzheimer's Disease · Deep learning · MRI images · Convolution Neural Networks

1 Introduction

Millions of people throughout the world suffer from the catastrophic neurological condition known as Alzheimer's Disease (AD), which is more common in older persons. The quality of life of affected individuals and their family is profoundly impacted by AD's progressive cognitive decline, memory loss, and behavioral changes. Since current medications are most effective when started early in the disease process, early identification of AD is essential for effective management and treatment. The symptoms of AD frequently arise gradually and are difficult to distinguish from normal aging, making early detection difficult.

Alzheimer's disease is a complicated, diverse condition that has many different effects on the brain. Amyloid plaques and neurofibrillary [2] tangles build up in the brain as a result of the disease, eventually killing brain cells and the gradual decline of

© The Author(s), under exclusive license to Springer Nature Switzerland AG 2024
S. Satheeskumaran et al. (Eds.): ICICSD 2023, CCIS 2122, pp. 29–42, 2024.
https://doi.org/10.1007/978-3-031-61298-5_3

cognitive ability. Currently, clinical assessment, cognitive testing, and brain imaging, particularly MRI, are used to make the diagnosis of AD.

However, because AD causes subtle changes in brain structure and function, it can be difficult to appropriately diagnose the condition using MRI pictures. The manual segmentation of different brain regions and the measurement of biomarkers, like cortical thickness or hippocampal volume [3], are the traditional approaches used to analyze MRI images. These techniques take a lot of time, are prone to mistakes and inter-rater variability, and demand a lot of knowledge, which prevents them from being widely used in clinical settings.

Recent developments in medical imaging, especially magnetic resonance imaging (MRI) [4], have made it possible to identify minor alterations in brain structure and function linked to AD. Traditional MRI image analysis techniques take a long time and need a lot of knowledge, which prevents them from being used frequently in clinical settings. Deep learning methods have recently come to light as a potential strategy for automated and precise AD identification from MRI scans.

The automatic learning of characteristics from complicated data, such as MRI images, is achieved via deep learning-based systems that employ neural networks with numerous layers. These strategies have achieved outstanding results in a number of domains, including speech recognition, natural language processing, and computer vision. Deep learning-based methods [20] for AD identification have the potential to automatically and accurately identify AD from MRI images, which could greatly enhance early detection and individualized treatment strategies.

These difficulties can be overcome by deep learning-based techniques, which enable precise and automated AD by automatically learning pertinent features from MRI scans detection. Using the use of massive datasets of MRI images, these techniques train neural networks [10] to recognize patterns and features that are challenging to spot using conventional techniques. The generated algorithms can then be applied to reliably forecast AD status from fresh MRI data.

Deep learning-based AD identification using MRI scans has showed encouraging results in recent research, displaying good accuracy and reliability in differentiating AD patients from healthy controls. These research have also demonstrated the potential for deep learning models [9] to recognize early AD indications before symptoms become evident, enabling early intervention and individualized treatment approaches.

In conclusion, deep learning-based methods present a viable route for enhancing MRI image-based early detection and treatment of AD. These methods are likely to become crucial weapons in the battle against this fatal illness as the field continues to develop.

An overview of deep learning-based techniques for AD identification from MRI images is given in this work. We go over the various deep learning architectures, [16] the difficulties with deep learning-based AD detection, and current research that has employed deep learning for AD identification using MRI scans. In order to improve early detection and treatment of AD, we examine future prospects and the possibility of deep learning-based techniques.

The rest of the paper is structured as follows. Section 2 is Literature survey that discusses the already existing projects and their disadvantages. It also discusses the proposed system structure of this project. Section 2.1, in this section we will discuss about the general issues identified from the previous work. Section 3, is proposed system in this section we will discuss about what is our proposed system and how it will work. Section 4 testing and validation of the system and design of test cases and scenarios. Section 5 conclusion of the system will be discussed.

2 Literature Review

Recent research has demonstrated that MRI pictures may reliably identify Alzheimer's Disease (AD) using deep learning techniques. In this review of the literature, we give a general overview of the state of the art in the study of detecting AD using deep learning on MRI scans.

A deep convolutional neural network (CNN) was employed in [13] to categorize AD patients and healthy controls based on MRI scans. They demonstrated that their model could identify between AD patients and healthy controls with excellent sensitivity and specificity, achieving an accuracy of 82%. In a different study, [17] used MRI scans to identify AD patients from healthy controls using a multi-channel CNN. They showed the promise of deep learning-based techniques for AD detection and attained an accuracy of 93.9%.

In [19] features were extracted from MRI images using a stacked auto encoder, and the accuracy for differentiating between AD patients and healthy controls was 86.7%. Similar to this, Liu et al.'s (2019) work obtained an accuracy of 92.2% when extracting features from MRI images using a 3D deep convolutional denoising auto encoder.

An accuracy of 92.6% was attained [6] classified AD patients and healthy controls using a transfer learning technique [14] and a pre-trained CNN. Deep learning-based methods for detecting AD from MRI scans have shown encouraging results, but they also have drawbacks. The requirement for huge and varied datasets to adequately train the models is one of the major problems. Another issue is the possibility of bias and overfitting, which may restrict the models' ability to be generalized to new datasets.

A CNN was employed to forecast the development of AD in people with mild cognitive impairment MCI MRI [11] pictures. They showed the potential of deep learning-based techniques for early intervention and individualized treatment by predicting the beginning of AD within three years with an accuracy of 79.2%.

A Holistic Approach Based on Machine Learning to Predict Patients' Clinical Courses Across the Alzheimer's Disease Spectrum [8]. The ML-based model [15] has an 86% accuracy rate in predicting whether MCI participants would become AD.

Autonomously divided brain MRI scans into various regions using a deep learning system, and then utilized these segmented pictures to forecast the existence of AD. On a collection of 568 MRI scans [21], they attained a classification accuracy of 85.1%. Neural network (CNN) to analyze MRI images and predict the progression of AD. They achieved a mean absolute error of 3.38 on a dataset of 630 MRI images.

Utilized a deep learning algorithm to examine brain MRI scans in [22] and forecast how AD will develop over time. With a dataset of 676 MRI scans, they were able to obtain a mean absolute error of 0. 035.

On a dataset of 145 MRI pictures in [12] employed a 3D CNN to analyze the images and predict the presence of AD with an accuracy of 86.4%. Long short-term memory (LSTM) network CNN [23] were used to analyze MRI images and forecast the alzhimer disease. A convolutional development of AD was employed in a work on a collection of 190 MRI pictures, they attained an accuracy of 87.1% in [24].

A deep learning method was utilized in [1] to analyze MRI scans and forecast the development of AD. In their method, a CNN was utilized to extract information from the MRI scans before a recurrent neural network (RNN) was used to forecast how AD will develop over time. With a dataset of 416 MRI scans, they were able to obtain a mean absolute error of 0.019.

A deep learning system was utilized in [7] to analyze MRI scans and identify the presence of AD. They used these segmented pictures to predict the presence of AD after training a CNN to automatically divide MRI images into various areas. An accuracy of 95.8% was attained on a dataset of 10,000 MRI scans. According to these findings, deep learning techniques MRI scans can be used to precisely predict the existence and development of Alzheimer's disease. In [5] used for the Fusiform Gyrus's functional connectivity during a face- matching task in people with mild cognitive impairment. Brain 129, 1113–1124.

Using fmri data [9] and deep learning convolutional neural networks, he had classified Alzheimer's disease; the work was published as an arXiv preprint. Overall, the research points to MRI as a helpful tool for detecting Alzheimer's disease and as a source of insightful data regarding the structure and operations of the brain.

2.1 Issues Identified

Alzheimer's disease is a degenerative neurological condition that impairs mental cognition. To stop the course of Alzheimer's disease and offer effective therapy, early detection is essential.

However, Alzheimer's disease is challenging to precisely diagnose, and many sufferers are not identified until the condition has already advanced severely.

As a result, the problem description for the early identification of Alzheimer's disease is the development of precise and trustworthy diagnostic tools and procedures. By identifying those who are at risk of getting Alzheimer's disease and giving them the right care and treatment, the development of such tools and approaches can assist those who already have the condition live better lives. The issue also calls for further investigation Given the intricacy of the condition and the difficulties in providing an accurate diagnosis, novel and creative techniques to Alzheimer's disease detection are required.

Accuracy of Diagnosis: Alzheimer's disease is difficult to diagnose because there is no single test that can do it. The use of multiple approaches, including MRI scanning, genetic testing, medical history, and cognitive tests, can provide a more accurate diagnosis of Alzheimer's disease. Since misdiagnosis can have serious repercussions, accurate diagnosis is essential. Underlying causes include genetics, environmental factors, and a person's lifestyle choices. Alzheimer's disease is a complex ailment. Further investigation is required in order to fully determine the condition's actual cause. It is challenging to create a cure or efficient treatments without a thorough grasp of the underlying reasons.

Alzheimer's disease progresses differently in each person and at varying rates. In order to create effective care and treatment strategies, it is essential to accurately forecast how quickly a patient's disease will progress. Patient safety is at danger due to the fact that Alzheimer's disease can seriously impair cognitive function and decision-making. Being able to anticipate and be ready for potential behavioral changes in Alzheimer's patients is crucial to providing top-notch care.

Despite growing public awareness of the disease, Alzheimer's is still a stigmatized condition. Alzheimer's patients, their families, and careers frequently come across prejudice and preconceptions. Raising public awareness and education can lessen stigma and make it easier for persons who have the condition to receive better support and care.

Ethical Issues: Alzheimer's disease is affecting an increasing proportion of the senior population, and caring for them can be taxing. Because of this, there are moral and ethical issues to think about when providing care for people with Alzheimer's disease, such as balancing the needs of the patient and the career and assessing their end-of-life choices. There are a number of other issues related to Alzheimer's disease that need to be addressed, in addition to the issue of accurate and early detection. These difficulties include lack of available effective treatments There is presently no cure for Alzheimer's disease, however there are certain drugs that can help manage the symptoms. Therefore, research into new treatments that can halt or delay the advancement of the illness is crucial.

Understanding the fundamental causes of Alzheimer's disease: Although the precise origins of Alzheimer's disease are still not entirely understood, it is thought that both genetic and environmental factors contribute to the condition's onset. Improved prevention and treatment methods might result from a deeper comprehension of the disease's root causes. Reducing stigma and enhancing care for those with Alzheimer's illness Alzheimer's illness still carries a lot of stigma, which can make it challenging for sufferers and their families to find support and care. Therefore, raising awareness and understanding of the condition and creating better support networks for those who have Alzheimer's disease and their families should be a priority.

Considering the financial effects of Alzheimer's: Alzheimer's disease has a major financial impact, both in terms of medical expenses and lost productivity. It may be possible to lower these expenses by creating more efficient preventative and treatment plans. Research funding must also be boosted in order to fully comprehend the disease's economic effects. Overall, overcoming the difficulties brought on by Alzheimer's disease necessitates a multifaceted strategy [25] that includes better diagnostic equipment, greater financing for research, and more efficient support networks for those who have the condition and their families.

3 Proposed System

The following procedures could be included in a proposed system for deep learning-based MRI image detection of Alzheimer's disease:

Data collection: The suggested system would start by collecting MRI scans from those who are at high risk of getting Alzheimer's disease. High-resolution T1-weighted pictures, DTI images, and functional MRI (fMRI) images are a few examples of these images.

Image Preparation To reduce noise and enhance picture quality, the acquired MRI images would undergo pre-processing. This would entail actions like motion correction, intensity normalization, and skull-stripping.

The features of the pre-processed MRI images would be learned using a deep learning model architecture, such as a convolutional neural network (CNN) or a recurrent neural network (RNN). While RNN can be used for sequence analysis, such as using fMRI data, CNN is used for 2D or 3D pictures.

Instruction of the Model: A sizable dataset of MRI images that have been previously processed and labelled to indicate whether the person has Alzheimer's disease or not would be used to train the deep learning model. In order to minimize the loss function, backpropagation and gradient descent would be used to optimize the model's parameters.

Testing and Evaluation: After the model has been trained, its performance will be tested on a different dataset. To do this, measures like sensitivity, specificity, accuracy, and area under the receiver operating characteristic curve would need to be calculated. (AUC-ROC).

Model Deployment: After the model has been verified, it may be put to use in a clinical environment to help with Alzheimer's disease diagnosis. The method could be used to help clinicians analyze MRI pictures and find early-stage Alzheimer's disease by being integrated into current clinical workflows.

In general, a deep learning model that can accurately identify people with Alzheimer's disease in their early stages would be developed and trained as part of the proposed system for Alzheimer's disease detection using MRI images. This approach could greatly enhance the effectiveness and accuracy of Alzheimer's disease diagnosis, thereby enhancing the quality of life for those living with the condition.

A plan for the creation or deployment of a new system inside a company or sector is often referred to as a proposed system. The proposed system might make use of cutting-edge tools, procedures, or tactics designed to solve a specific problem or boost productivity.

Research, analysis, design, development, testing, and deployment are typical steps that the proposed system goes through. The suggested system's target problem is located during the research and analysis phase, and the solution's specifications are established.

The architecture of the proposed system is established during the design phase while accounting for the numerous components and functions needed. The system is constructed during the development stage using the design specifications. The system is assessed during testing for its usability, effectiveness, security, and dependability. The system can be introduced to the company or sector for use by stakeholders after testing is complete.

In general, a suggested system needs rigorous planning, design, and execution to be successful because it is meant to solve a specific issue, lower costs, or increase efficiency. Convolutional Neural Networks (CNNs) are a class of deep learning models that are frequently applied to speech recognition, natural language processing, image and video analysis, and other tasks. The key characteristic of CNNs is their collection of convolutional layers, each of which is intended to automatically recognize and extract pertinent characteristics from input images.

Input, convolutional, pooling, fully connected, and output layers are among the layers that make up CNNs. The raw image data is fed into the network at the input layer. The convolutional layer extracts important features like edges, corners, and textures by scanning the input image with a collection of learnable filters. The output of the convolutional layer is subsequently downscaled using the pooling layer, which lowers the number of parameters in the increases the computational effectiveness of the model. The output of the pooling layer is passed on to the fully connected layer, which converts it into a collection of probabilities showing the chance that the input image belongs to certain classes. The ultimate categorization result is provided by the output layer.

The main benefit of CNNs is its capacity to learn and extract pertinent features automatically from input images without the need for manual feature engineering. Due to the complexity and difficulty of manually defining the features of the input photos, they complex tasks such as image captioning and video analysis.

On a variety of image classification tasks, such as object recognition, face recognition, and medical image analysis, CNNs have been utilized to reach state-of-the-art performance. The use of them is also found in learning models, such as recurrent neural networks, to perform more complex tasks such as image captioning and video analysis (Fig. 1).

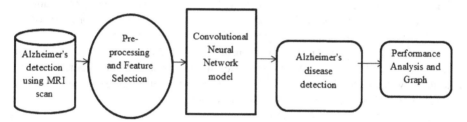

Fig. 1. Architecture diagram for the Alzheimer's disease detection

The architecture for deep learning-based Alzheimer's disease detection utilizing MRI images typically consists of the following elements:

The pre-processed MRI pictures are supplied into the deep learning model at the input layer. The images could be in T1-weighted images, DTI images, or fMRI images, among other formats, and they could be in 2D or 3D.

Continuum Layers: The fundamental units of deep learning models for image processing are convolutional layers. The important features from the input photos, like edges, corners, and textures, are extracted using filters. A series of feature maps are produced by the convolutional layers and are then transferred to the following layer.

Pooling Layers: To down sample the feature maps and lower the model's parameter count, pooling layers are often applied after the convolutional layers. Max pooling and average pooling are two popular varieties of pooling layers. Activation Functions: To provide nonlinearity to the deep learning model, activation functions are used. ReLU, sigmoid, and tanh are the activation functions that are most frequently utilized.

Fully Connected Layers: The final classification of the input photos is carried out using fully connected layers. They convert the output of the convolutional and pooling layers into a collection of probabilities that show how likely it is that the input images will fall into certain groups.

The deep learning model's output layer is where the final classification of the input photos is made. The output layer may have two nodes for Alzheimer's disease and normal control in order to detect the disease.

Loss Function: Convolutional layers, pooling layers, fully connected layers, activation functions, and an output layer are frequently combined for identifying Alzheimer's disease using MRI images and deep learning. Depending on the specific requirements of the situation and the data at hand, the design can be adjusted and optimized.

When detecting Alzheimer's disease using MRI scans and deep learning, convolutional layers, pooling layers, fully connected layers, activation functions, and an output layer are typically integrated. The design can be modified and optimized based on the specific requirements of the problem and the available data.

In conclusion, the deep learning-based method for diagnosing Alzheimer's disease from MRI scans holds great promise for improving the accuracy and efficacy of the diagnosis. By exploiting the capabilities of deep learning algorithms, the system can efficiently examine MRI images and identify minute changes in brain structure that are symptomatic of Alzheimer's disease.

The system's ability to detect and diagnose Alzheimer's disease early can have a big impact on patient outcomes and total healthcare costs. Early intervention enables patients to receive the care they require when they require it, which can slow the progression of the illness and improve the patient's quality of life.

The recommended system has to be further validated and tested on a larger and more diversified dataset in order to guarantee its dependability and generalizability. Additionally, the technology must be integrated with the current healthcare system to guarantee its wide acceptance and patient accessibility.

The steps required for execution of our system:

1. First we will take the dataset from Kaggle and by using data augmentation we expanded the dataset for multiple input testing.
2. By using convolutional neural networks(CNN), we created a model for testing the dataset
3. Web application is used for creating a web page to upload the specified data.
4. Web page is designed in such a way that after executing the html code, it directs the application into a login page.
5. By entering the username and password(login credentials), we submit them by clicking on submit button which is on bottom center of the login page
6. A preview page is displayed where it shows a title "Alzheimer"
7. and it is used for uploading the desired image for MRI scan image of brain.
8. By submitting the image it directs the page into a result page where it shows the image followed an output of the stage that particular person is suffering from.
9. There will be a performance analysis button which is shown on top right corner of login page where we can see the accuracy and confusion matrix of our model.

4   Result and Discussion

In this paper, we suggest a deep learning-based method for spotting early signs of Alzheimer's disease. The Convolutional Neural Network was the deep learning method employed. A GUI has been built so that users may easily upload an MRI scan of their brain, and the result will be displayed when the image has been uploaded.

Here are a few screenshots demonstrating the absence and potential severity of the condition. Following image upload (Fig. 2).

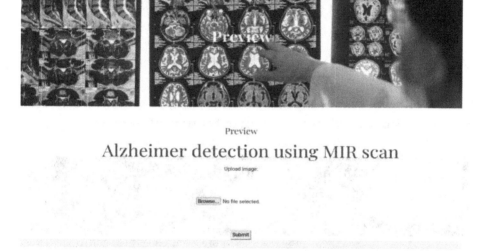

Preview

Alzheimer detection using MIR scan

Upload image:

Browse... No file selected.

Submit

Fig. 2. Screenshot for the uploading of MRI brain images.

After successful login into the web site the preview page will be opened in that preview page, we have a browse option which is useful for the upload of images from our specific location. First, we will save some of MRI images for the testing purpose which is used for the testing of our model with that images. When we click on brows option, we have to select an image from our device then we have to upload the image into out model then it will show the specific image with the submit button at last we have to click on submit. After submitting the image, it will give results of specified severity of the disease (Figs. 3 and 4).

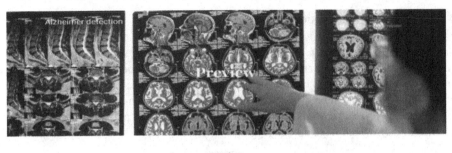

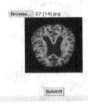

Fig. 3. Screenshot of after uploading the image into the webpage

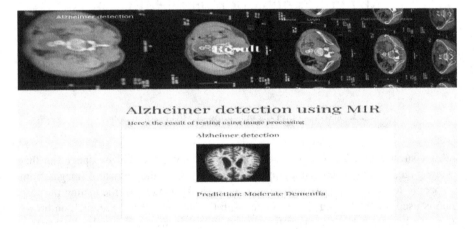

Fig. 4. Screenshot showing presence of Alzheimer's disease of moderate demented severity.

4.1 Result Analysis

After successful execution of our system the results are as follows:

Confusion matrix with accuracy: A confusion matrix is a table that is used to evaluate the performance of a classification model. It is a matrix that compares the predicted output of a model with the actual output. The matrix displays the number of true positives (TP), true negatives (TN), false positives (FP), and false negatives (FN) produced by a model.

So below is the representation of accuracy by using confusion matrix for our system (Fig. 5).

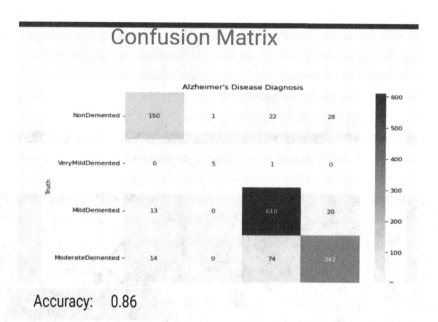

Fig. 5. Confusion Matrix of our proposed System for Accuracy

Here the accuracy of the proposed system as in Fig. 8 is 86%. The below chart represents that classification of all types of dementia based on the given input dataset.

Now let us compare our model accuracy with the other models.

Inceptionv3 is a type of deep learning classification model which is used for the classification of images but when we compare both models accuracy our model which is detection of Alzheimer's with CNN has given more accuracy that is 86% than the inceptionv3 model. Where inceptionv3 model given only 76% accuracy only as in Fig. 6.

Now let us compare deep leering model accuracy with the machine learning model which is SVM support vector machine (Fig. 7).

Here we can observe that the accuracy is very low which is of 31% with the SVM so this model not suitable for the classification of our images. When compared to our proposed system with CNN model it has given more accuracy than SVM. SVM stands for Support Vector Machines, which is a type of supervised machine learning algorithm used for classification and regression analysis. In SVM, a hyperplane is used to separate the data points into different classes based on their features.

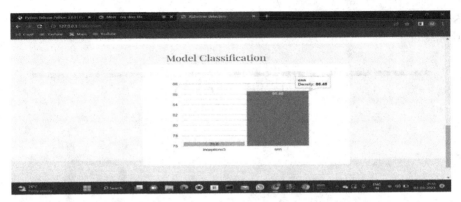

Fig. 6. Inceptionv3model vs our proposed CNN model

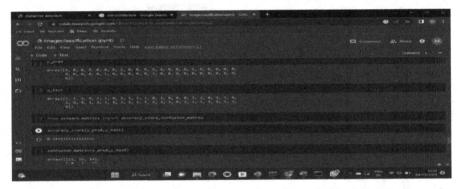

Fig. 7. SVM Vs Deep Learning CNN model

5 Conclusion

In this Work, we used the basic Convolutional Neural Network (CNN) architecture model classify Alzheimer's from magnetic resonance imaging (MRI) scans images. Convolutional Neural Network (CNN) architecture model is used to avoid the expensive training from scratch and to get higher efficiency with limited number of dataset. The proposed work was able to give a Good accuracy where training accuracy is 86.34% and validation accuracy is 86.45% on the test data with very small misclassifications on normal and very mild demented.

In this study, we applied the core Convolutional Neural Network (CNN) architecture model on images from MRI scans to classify Alzheimer's illness.

Convolutional neural networks (CNNs) are employed as part of architectural models to increase efficiency with fewer datasets and prevent expensive retraining.

When compared to other detection systems like Inceptionv3, our deep learning-based system's accuracy was high, whereas other methods, such machine learning techniques, produced results that were far less accurate than those of our system.

References

1. Jack, C.R., Jr., et al.: NIA-AA research framework: toward a biological definition of Alzheimer's disease. Alzheimer's Dementia **14**(4), 535–562 (2018)
2. Bhagwat, N.: Prognostic applications for Alzheimer's disease using magnetic resonance imaging and machine learning, doctoral dissertation, graduate programme in biomedical engineering, University of Toronto, Toronto, Ontario, Canada (2018)
3. Jack, C.R., Jr., et al.: Age, sex, and APOE ε4 effects on memory, brain structure, and β-Amyloid across the adult life span. JAMA Neurol. **72**(5), 511–519 (2015)
4. Chaddad, A., Desrosiers, C., Niazi, T.: Deep radiomic analysis of MRI related to Alzheimer's disease. IEEE Acces **6**, 58213–58221 (2018)
5. Sabbagh, M., Shi, J., Paul, G., Jackson, R., Mehta, D.: Why do treatment studies for Alzheimer's disease always coming up empty? An overview of discontinued drugs from 2010 to 2015. Expert Opin. Investig. Drugs **26**(6), 735–739 (2017)
6. Chen, R., Shi, L., Yan, S., Shaik, N., Li, X., Haleshappa, R.A.: Early Alzheimer's Disease Diagnosis Using Machine Learning and Image Analysis. Front Public Health, 2627–2635(2022)
7. Van Someren, E.J.W., et al.: Medial temporal lobe atrophy relates more strongly to sleep-wake rhythm fragmentation than to age or any other known risk. Neurobiol. Learn Mem. **160**, 132–138 (2019)
8. Vatanabe, I.P., Manzine, P.R., Cominetti, M.R.: Historic concepts of dementia and Alzheimer's disease: from ancient times to the present. Rev. Neurol. (Paris) **176**(3), 140–147 (2020)
9. Ahmed, S., et al.: Ensembles of patch-based classifiers for diagnosis of Alzheimer diseases. IEEE Access **7**, 73373–73383 (2019)
10. Turk, M., Tofighi, G., Sarraf, S.: Utilizing fMRI data and deep learning convolutional neural networks to classify Alzheimer's illness. preprint from arXiv: 1603.08631 (2016)
11. Liu, M., et al.: DAutomatic classification of Alzheimer's disease and mild cognitive impairment using a deep convolutional neural network based on T2-weighted MRI. J. Alzheimer's Dis. **73**(4), 1469–1479 (2020)
12. Dolz, J., Desrosiers, C., Ayed, I.B.: 3D fully convolutional networks for subcortical segmentation in MRI: a large-scale study. Neuro Image **170**, 456–470 (2017)
13. Tofighi, G., Sarraf, S.: fMRI data is used in a deep learning-based pipeline to identify Alzheimer's disease. In: Medical Image Computing and Computer-Assisted Intervention at the International Conference, pp. 475–483. Springer, Cham (2016)
14. Huang, J., et al.: Deep learning-based discriminative analysis of multimodal imaging data for early-onset Alzheimer's disease. Ageing Neurosci. Front. **10**, 385 (2018)
15. Zhang, D., et al.: Using a deep convolutional neural network based on T2- weighted MRI, Alzheimer's disease and moderate cognitive impairment are automatically classified. J. Alzheimer's Dis. **73**(4), 1469–1479 (2020)
16. Sivaswamy, J., Jayasree, R.S., Gopinath, K.: Convolutional neural network with transfer learning for Alzheimer's disease detection using structural MRI. IEEE Access **8**, 8801–8811 (2020)
17. Gao, Y., Sarraf, S.: Alzheimer's disease classification based on brain MRI data and deep learning. In: Presented Paper at the SPIE Medical Imaging Conference in San Diego, California (2016)
18. Zhao, Y., Yang, Y., Zhu, X., Li, K., Chen, Z.: Alzheimer's disease diagnosis utilizing a convolutional neural network-based deep learning method. J. Med. Syst. **43**(8), 239–245 (2019)
19. Huang, L., Wei, W., Zhao, X., Xie, S.: A 3D deep learning method based on various MRI modalities for classifying Alzheimer's disease. Front. Neurosci. **13**, 1008–1023 (2019)

20. Zhang, Y., Su, J., Mo, Y., Qin, J., Wang, S.: A 3D CNN-based multi-level feature extraction technique for the diagnosis of Alzheimer's disease based on structural MRI and sMRI data. Front. Neurosci. **13**, 339–356 (2019)
21. Liu, S., et al.: Learning multimodal neuroimaging feature for several Alzheimer's disease classifications. IEEE J. Biomed. Health Inform. **24**(1), 26–34 (2020)
22. Wang, Y., et al.: Using MRI scans, deep feature learning is used to diagnose Alzheimer's disease. Front. Aging Neurosci. **12**, 285–316 (2020)
23. Wang, Q., et al.: Using deep learning, we can jointly diagnose and forecast the prognosis of Alzheimer's disease based on inadequate multimodality data. Med. Image Anal. **62**, 101674–101690 (2020)
24. Shi, Y., et al.: For the purpose of diagnosing Alzheimer's disease, hierarchical multimodal fusion of structural and functional brain networks is used. IEEE Trans. Med. Imag. **39**(9), 3006–3016 (2019)
25. Zhang, X., Su, Y., Li, Y., Xiao, Y., Zhu, M., Li, K.: Alzheimer's disease diagnosis based on T1-weighted magnetic resonance imaging and three-dimensional deep convolutional neural networks. Med. Sci. Monit. **26**, 921837–921867 (2020)

A Comparison Study of Cyberbullying Detection Using Various Machine Learning Algorithms

Chaitra Sai Jalda[1], Uday Bhaskar Polimetla[2], Ashok Kumar Nanda[1(✉)], and Shivangi Nanda[3]

[1] Department of CSE, B V Raju Institute of Technology, Narsapur, Medak 502313, Telangana, India
ashokkumarnanda@yahoo.com
[2] Nexgenai Solutions Private Limited, Hyderabad 500016, Telangana, India
[3] School of Electronics Engineering, VIT-AP University, Amaravathi 522237, Andhra Pradesh, India

Abstract. Cyberbullying is one of the most notable current social media issues. Cyberbullying is a type of bullying or harassment that occurs online. The prevalence of cyberbullying is growing as the digital age and technology advances. Cyber-threats are now a significant problem that affects students, particularly teenagers. The victims of bullying may experience negative effects on their emotional health. Additionally, awareness has grown because of some suicide incidents. Bullying occurs most frequently on networks with direct messaging. The availability of tools that can automatically detect potential behaviors categorized as cyberbullying can be useful. In this paper, we have compared the accuracy results of various machine learning algorithms for detecting, identifying, and classifying the cyberbullying texts or sentences on various social media sites like Twitter that could possibly lead to a cyberbullying episode.

Keywords: Classifying · Cyberbullying · Detecting · Identifying · Machine Learning · Vectorizer

1 Introduction

Social networks are popular as a main source for spreading messages to other people. However, it also provides an opportunity to create harmful activities. There is numerous evidence showing that messaging can introduce a very concerning problem, namely cyberbullying. To detect cyberbullying, we use machine learning classifiers for analysing Twitter datasets and implement them on a website that can detect bullying text. As a result, cyber-hate crime has increased dramatically in recent years. Social media has a lot of benefits, but they cannot outweigh the drawbacks. Cyberbullying is one of the most notable recent social media issues. Social networking has a lot of benefits, but it also has some serious drawbacks. This medium is used by malicious people to commit unethical and dishonest behaviours that hurt other people's feelings and damage their reputations.

© The Author(s), under exclusive license to Springer Nature Switzerland AG 2024
S. Satheeskumaran et al. (Eds.): ICICSD 2023, CCIS 2122, pp. 43–54, 2024.
https://doi.org/10.1007/978-3-031-61298-5_4

Cyberbullying, also known as cyber harassment, is a type of bullying or harassment that occurs online. There are many different types of cyberbullying, such as cyber harassment. As the digital era and technology have advanced, cyberbullying has become more prevalent, particularly among young people. Bullying occurs most frequently on networks with direct messaging. Cyberbullying, or cyber harassment, is the act of bothering internet users via technological means. The victims of bullying may experience negative effects on their emotional health. In this respect, a cyberbullying detection model based on machine learning can ascertain whether or not a communication is related to cyberbullying. Many machine learning techniques, such as Naive Bayes, support vector machines, decision trees, and others, are included in the proposed cyberbullying detection model.

The problem of detecting cyberbullying through machine learning involves identifying instances of cyberbullying in digital communication and classifying them as either cyberbullying or not. This can be a challenging task, as cyberbullying can often be subtle and may not always be explicitly apparent from the content of the communication. It may also involve detecting patterns of behaviour that are indicative of cyberbullying, such as repeated or persistent aggression or intimidation. The goal here is to develop a machine learning model using a Twitter dataset that can accurately detect instances of cyberbullying in the given text. Machine learning models will be evaluated based on their ability to correctly identify instances of cyberbullying. This is measured through accuracy.

In this paper, a literature review of previously proposed papers is presented in Sect. 2. The methodology of the proposed system is discussed in Sect. 3. Performance analysis is presented in Sect. 4. Sections 5 and 6 represent conclusions and future enhancements.

2   Literature Survey

In the past few years, there has been a lot of research that has been proposed in the field of cyberbullying, maybe text-based or image-based bullying.

The Naive Bayes algorithm produced significantly superior results in studies on the detection and classification of cyberbullying. When trained on Facebook comments in several categories, such as shame, harassment, and racism, the Multinomial Naive Bayes algorithm provided an accuracy of 88% [3], according to Arnisha Akhter, Uzzal K. Acharjee, and Md. Masbaul A. Polash. The percentage of each type of bully was also calculated. However, a lot of other things weren't considered, such as gender or country. Muskan Patidar, Mahak Lathi, and Prof. Yamini Barge [4] gave a solution to a different dataset that consists of posts related to cyberbullying on Twitter. It was observed that Naïve Bayes gave the best accuracy in being able to identify a comment as a bullied or non-bullied statement. In all the 4 types of Naïve Bayes evaluated (bi grams, tri grams, n grams, uni grams), N grams gave the best accuracy of 66.3%.

To evaluate the correctness of the results, some researchers used fundamental machine learning techniques against various datasets. A machine learning-based cyberbullying detection model to check if the communication was bullying or not was proposed by A. Varsha Reddy, Gugulothu Kalpana, and N. Satish Kumar [5]. Logistic regression paired with a separate characteristic vector-like count vectorizer outperformed all other combinations. This combination resulted in an accuracy of 80.01%. Similarly, Prof. Jaya Jeswani, Vedant Bhardwaj, and Bhavin Jain [14] proposed an initiative that seeks to meet the demand for locating unfavourable tweets that encourage hate speech. This study suggested a system that processes user-supplied custom input in a Jupyter notebook to determine whether the input string is nasty or not; excellent results were obtained through logistic regression and decision trees after analysing several ML algorithms. But because there was an imbalanced label between hatred and no hate, a robust, advanced manual tagging mechanism was required. Chahat Raj, Ayush Agarwal, and Gnana Bharathy, et al. [10], performed research to assess the impact of feature extraction and word embedding strategies on algorithmic performance. Bidirectional neural networks and attention models produced high classification results, according to key findings from this study. With neural network models, Global Vectors (GloVe) functioned more effectively.

Comparing hybrid models to a single algorithm, they are more effective. Some methods for detecting cyberbullying have been developed utilising hybrid algorithms. Dr. Vijayakumar V, Dr. Hari Prasad D, and Adolf P [9] A hybrid deep neural network model was created in this study to identify cyberbullying, including text and pictures. Using the Kears Functional API, the model was created using a convolutional neural network (CNN) and long short-term memory (LSTM) combo. After training the model, the image-based forecast had an accuracy of 86%, and the text-based prediction had an accuracy of 85%. Theyazn H. H. Aldhyani and Mosleh Hmoud Al-Adhaileh [8] published a paper with the primary goal of creating a system for detecting cyberbullying using a hybrid deep learning design made up of convolutional neural networks coupled with bidirectional long short-term memory networks (CNN-BiLSTM) and single BiLSTM models. With a detection rate of 94%, both classifiers performed admirably in the binary classification dataset (aggressive or non-aggressive bullying). BiLSTM performed better than the combined CNN-BiLSTM classifier for the multiclass dataset, reaching 99% accuracy. Shardul Suryawanshi, Bharathi Raja Chakravarthi, Mihael Arcan, and Paul Buitelaar [12] proposed a method for classifying inappropriate meme content based on the graphics and words that go along with it. To train and test a multimodal classification system for identifying objectionable memes, the MultiOFF dataset was employed. Better outcomes than evaluating independently are demonstrated when stacked LSTM, BiLSTM, and CNN multi-models are employed.

Some of the unique approaches to the detection of cyberbullying include a temporal graph-based cyberbullying detection framework (TGBully) by Suyu Ge Lu Cheng Huan Liu [2], which focuses on user interaction. When the dataset was oversampled, this suggested solution achieved an accuracy of 79%, and when it was not, it gave an accuracy of 82%. A significant finding was that user interaction on social media is represented by the relationships between their posted comments. Another approach is by Hassan Awad Hassan Al-Sukhni and Azuan Bin Ahmad [6], in which a technique to detect cyber terrorists suspected activities over the net is proposed by combining the Krill Herd and Simulated Annealing algorithms. To maximise the accuracy rate, this paper introduces three new degrees of categorization: low, high, and interleave. According to the experiment, this method outperformed the benchmark work, producing an interleave level accuracy rate of 73.01%.

Yazid et al., 2022 [11]. This study used the fuzzy Delphi method to revalidate the cyber aggressor scale for the Malaysian context. The fuzzy Delphi approach was used to study expert comments to assess the applicability of the indicators. Experts concur that using the fuzzy Delphi method to validate the items in the suggested cyberbullying scale is an efficient methodology. The analyses' findings supported the CYB-AGG scale' psychometric robustness. The researcher exclusively used Malaysian experts in this study, which had its own constraints.

Another of the best machine learning methods is neural networks. They are also effective at detecting cyberbullying, according to several studies. Maral Dadvar and Kai Eckert released an article [1] on the use of DNN-based models to identify instances of cyberbullying on social media sites. In this study, machine learning models using DNN-based models in combination with other transfer learning techniques produced the best results on a dataset from YouTube. 76% accuracy was achieved using this method. Belal Abdullah Hezam Murshed and Jemal Abawajy's paper [7], which provided a useful strategy to improve the effectiveness of detecting cyberbullying occurrences, is another beneficial piece of work. The DEA optimisation and the Elman-type RNN were coupled to generate effective parameter tuning.

Mr. Shivraj Sunil Marathe and Prof. Kavita P. Shirsat [13] examined numerous strategies that have been created to identify cyberbullying on various social media platforms. It was determined that most studies used classifier approaches like SVM. To identify every kind of cyberbullying, a novel approach must be developed. When multiple features are used to detect cyberbullying, performance is greater in terms of accuracy, precision, recall, etc. than when utilising user-based, content-based, or contextual features separately.

Several studies related to cyberbullying analysis and detection using text mining by classifying conversations or postings using unsupervised learning and labelling using N-grams and TF-IDF. In a supervised machine learning approach, they collected YouTube comments, labelled them manually, and implemented various binary and multiclass classifications.

3 Methodology

This section consists of the proposed system, architecture, modules, and UML diagrams of the proposed solution for cyberbullying detection.

3.1 Proposed System

Cyberbullying victims may experience anxiety, sadness, low self-esteem, and even suicidal thoughts as a result of the abuse, which can have major and long-lasting impacts on their mental health and wellbeing. Cyberbullying may result in serious legal and societal repercussions. The ability to identify and stop cyberbullying can help safeguard the welfare and safety of people and communities, as well as foster an accepting and positive online environment.

Therefore, we provide a paradigm in which a set of cyberbullying tweets is used to train machine learning algorithms (Naive Bayes, Logistic Regression, Decision Tree,

AdaBoost, and Random Forest). This dataset, which has over 48,000 entries, was taken from the Kaggle website. This dataset contains tweets that have been classified as either bullying or non-bullying. Gender, ethnicity, age, and other relevant hatred are contributing variables to bullying in tweets. Accuracy results are produced after analysis of the algorithms. The algorithm that provides us with optimal accuracy is also deployed on a website that classifies texts and determines whether or not they are bullying.

3.2 Architecture

The architecture of the proposed solution for detecting cyberbullying using the analysis of various machine learning algorithms is depicted in Fig. 1. The steps include pre-processing and feature selection. The role of feature selection is to reduce the feature dimensionality by removing irrelevant features and selecting only the most important features that are used in the present predictive model. The most important stage that follows the feature selection process is the classification stage. In the current model, various machine learning algorithms are analysed using a Twitter dataset to check their performance, and the optimal machine learning algorithm is further implemented on a website where a given text can be predicted to be bullying or non-bullying.

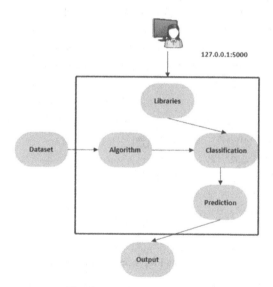

Fig. 1. Architecture Diagram

3.3 Modules

We mainly have three modules in this project. Such as a. Module Training Module; b. Home Module; and c. Result Module.

3.3.1 Module Training Model

In this module, the data set is collected from the Kaggle website, and the data is pre-processed and converted using a count vectorizer. The testing training dataset is then divided. Different algorithms are compared to detect an algorithm that can give us the best accuracy for the dataset taken. A pictorial representation of the flow of the proposed model is shown in Fig. 2 below.

That algorithm is initialised, and feature labels are fitted to the Logistic Regression Algorithm (optimal). Then, give an IP address for accessing the website. Now, we need to identify that IP address for accessing the website and open it in a new browser. To know whether the message is a cyberbullying text or not, we need to type in the space provided by the website, and then it will output whether the text is cyberbullying or not.

3.3.2 Home Module

The Home module has HTML programming where an IP address is connected, messages are managed by communicating with clients, and a trained model is loaded into the space given on the website to determine whether the given text is bullying text or not. So to identify the text, we will type in the space given in the home page formed by using HTML programming, and the result, whether it is bullying or not, is displayed in the result page, which is explained in the next module in detail.

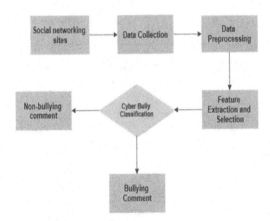

Fig. 2. Data Preprocessing Module

3.3.3 Result Module

The result module is also an HTML framework that is connected to an IP and gives the result analysis of the machine learning model from home. The result of the text messages given on the home module page is the conclusion of the whole paper. This result page is also formed by using HTML programming. Both the home and result pages are formed by HTML programming. The messages given on the home page and the results displayed on the result page are the main parts of the project. The HTML programming must be done carefully in order to avoid further complications in the project.

3.4 UML Diagrams

This subsection consists of UML diagrams of the proposed solution.

3.4.1 Data Flow Diagram

The diagram below Fig. 3 represents a data flow diagram (DFD), which shows how information moves through the system and how it is used to carry out various tasks. This DFD diagram clearly shows the scope and bounds of the system. It is a picture of how information moves through a system or a process. It also gives insight into the inputs and outputs of each entity.

3.4.2 Sequence Diagram

The most popular type of interaction diagram is a sequence diagram. Below the depicted picture, Fig. 4 is the sequence diagram of the proposed model. A system's interacting behaviour is depicted by an interaction diagram. We utilise many forms of interaction diagrams to capture different elements and components of interaction in a system.

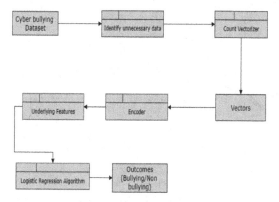

Fig. 3. Data Flow Diagram

3.4.3 Use Case Diagram

A use case is a description of a series of actions that a system takes to produce a valuable result for an actor. Use case diagrams address the system's static use case perspective. Use case diagrams to show what a system performs, not how it will be carried out. Additionally, use cases can be arranged using relationships between them that determine generalisation, inclusion, and extension. Figure 5 represents the use-case diagram of the proposed system.

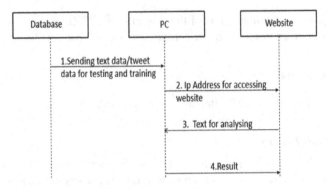

Fig. 4. Sequence Diagram

3.4.4 Activity Diagram

An activity diagram is a form of flowchart that shows how a system or activity's workflow or process flows. The sequence of required actions is shown in Fig. 6, together with the decision points, branching, and flow of control from beginning to end. Activity diagrams can be used to describe the steps in a use case, the flow of control in a business process, or the interactions between system objects. They are used to model the dynamic aspects of a system.

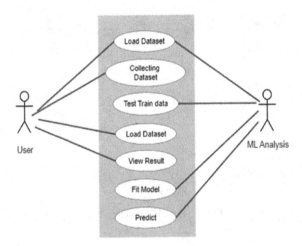

Fig. 5. Use case Diagram

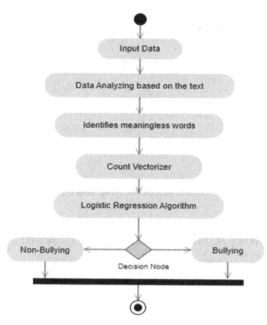

Fig. 6. Sequence Diagram

4 Performance Analysis

The below Fig. 11 shows 5 different algorithms, namely, Naïve Bayes, Logistic Regression, Decision Tree, AdaBoost, and Random Forest, that were trained on a Twitter dataset, which resulted in the following (Fig. 7):

Bullying Message

During implementation, we tested bullying and non-bullying words with the above-mentioned algorithms. Logistic Regression provides more accuracy than other algorithms. Our proposed model will identify whether there are any bullying words in the given sentence. The below-mentioned Fig. 8, Fig. 9 and Fig. 10 are the results showing whether the given sentence has bullying words or not.

Non-bullying Message

The two texts are provided to determine whether they are bullying or non-bullying messages.

From the graph, we have concluded that Logistic Regression provides more accurate results than other algorithms like Naïve Bayes, Decision Tree, AdaBoost, and Random Forest.

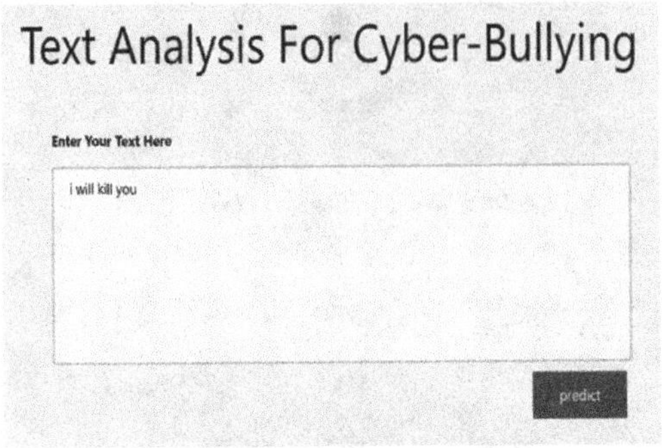

Text Analysis For Cyber-Bullying

Enter Your Text Here

i will kill you

predict

Fig. 7. Screenshot of Bullying Text (home)

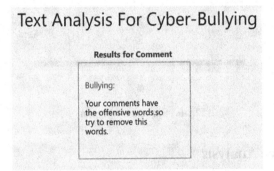

Fig. 8. Screenshot of Bullying Text (result)

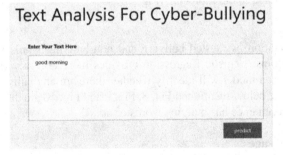

Fig. 9. Screenshot of Non-Bullying Text (home)

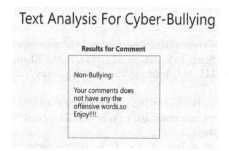

Fig. 10. Screenshot of Non-Bullying Text (result)

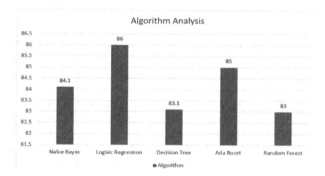

Fig. 11. Graph Comparing the Accuracies.

5 Conclusion

Most people are using bullying words on social networks like Twitter. We can reduce online harassment and abuse by using machine learning algorithms. From the results, it is concluded that Logic Regression is a more effective algorithm to identify bullying words. Providing a platform to further assess whether a communication is bullying or not. This might be a potent tool to combat online abuse and harassment. Websites can rapidly and accurately detect instances of cyberbullying and take the necessary precautions to protect their users by utilising machine learning algorithms. Websites may assist in reducing the negative impacts of cyberbullying on people and communities by putting in place efficient techniques for detecting it and fostering a safer and more welcoming online environment.

6 Future Work

Here, we performed analysis only on the content of tweets; we could not perform the analysis in relation to the user's behaviour. In future, it can be extended to analyse different types of media, such as images, video, and audio. This can also be extended in such a way that it can automatically detect bullying messages in a live chat, i.e., in social media networks, which can save many people from danger.

References

1. Dadvar, M., Eckert, K.: Cyberbullying detection in social networks using deep learning based models. In: Song, M., Song, I.-Y., Kotsis, G., Tjoa, A.M., Khalil, I. (eds.) DaWaK 2020. LNCS, vol. 12393, pp. 245–255. Springer, Cham (2020). https://doi.org/10.1007/978-3-030-59065-9_20

2. Ge, S., Cheng, L., Liun , H.: Improving cyberbullying detection with user interaction. In: Proceedings of the Web Conference 2021 (WWW 2021), pp. 496–506 (2021). https://doi.org/10.1145/3442381.3449828

3. Akhter, A., Acharjee, U.K., Polash, M.M.A.: Cyber bullying detection and classification using multinomial naïve bayes and fuzzy logic. Int. J. Math. Sci. Comput. 5(4), 1–12 (2019)

4. Patidar, M., Lathi, M., Jain, M., Dhakad, M., Barge, Y.: Cyber bullying detection for twitter using ML classification algorithms. Int. J. Res. Appl. Sci. Eng. Technol. 9(11), 24–29 (2021). ISSN: 2321-9653

5. Varsha Reddy, A., Kalpana, G., Satish Kumar, N., Dheeraj, S.: Cyber bullying text detection using machine learning. Int. J. Res. Appl. Sci. Eng. Technol. 10(6), 2868–2872 (2022)

6. Al-Sukhni, H.A.H., Ahmad, A.B., Saudi, M.M., Alwi, N.H.M.: Cyber terrorist detection by using integration of krill herd and simulated annealing algorithms. IJACSA 10, 317–323 (2019)

7. Murshed, B.A.H., Abawajy, J., Mallappa, S., Saif, M.A.N., Al-Ariki, H.D.E.: DEA-RNN: a hybrid deep learning approach for cyberbullying detection in Twitter social media platform. IEEE Access 10, 25857–25871 (2022)

8. Aldhyani, T.H.H., Al-Adhaileh, M.H., Alsubari, S.N.: Cyberbullying identification system based deep learning algorithms. Electronics 11, 3273 (2022). https://doi.org/10.3390/electronics11203273

9. Vijayakumar, V., Hari Prasad, D., Adolf, P.: Multimodal cyberbullying detection using hybrid deep learning algorithms. Int. J. Appl. Eng. Res. 16, 568–574 (2021). ISSN 0973-4562

10. Raj, C., Agarwal, A., Bharathy, G., Narayan, B., Prasad, M.: Cyberbullying detection, hybrid models based on machine learning and natural language processing. Electronics 10(22), 2810 (2021)

11. Yazid, Z.N.A., Bakar, A.A., Hashim, J.N., Aziz, N.N.A.: Revalidating adolescent cyberbullying scale using fuzzy Delphi approach. Int. J. Acad. Res. Bus. Soc. Sci. 12(9), 844–855 (2022)

12. Suryawanshi, S., Chakravarthi, B.R., Arcan, M., Buitelaar, P.: Multimodal meme dataset (MultiOFF) for identifying offensive content in image and text. In: Proceedings of the Second Workshop on Trolling, Aggression and Cyberbullying, pp 32–41 (2020)

13. Marathe, S.S., Shirsat, K.P.: Approaches for mining YouTube videos metadata in cyber bullying detection. Int. J. Eng. Res. Technol. 4(05), (2015). ISSN:2278-0181

14. Jeswani, J., Bhardwaj, V., Jain, B., Kohli, B.S.: Negative sentiment analysis: hate speech detection and cyber bullying. Int. J. Res. Appl. Sci. Eng. Technol. 10(5), 911–917 (2022). ISSN:2321-9653

State of the Art Analysis of Word Sense Disambiguation

Madhuri Karnik, Vaishali Mishra, Vidya Gaikwad, Disha Wankhede, Priya Chougale,
Vaishnavi Pophale[✉], Arpit Zope, and Chirag Maski

Vishwakarma Institute of Information Technology, Pune, India
`vaishnavi.22010473@viit.ac.in`

Abstract. Word Sense Disambiguation (WSD) is a computational activity in natural language processing (NLP) that aims to determine the meaning of a word in its context by choosing the most appropriate meaning or definition from a predefined set of possibilities. WSD is essential for many applications such as machine translation, information retrieval, text classification, question answering, summarization, sentiment analysis, and word processing. WSD is a difficult task, especially in cases where words have multiple meanings or the context is ambiguous, which can lead to difficulties in choosing the correct meaning of a word. The WSD has applied many strategies to different datasets and corpora. This study used a knowledge-based approach and machine learning techniques to classify WSD algorithms. Each category is examined in-depth, along with an explanation of its associated algorithms. The paper reviews various approaches, and resources, used in WSD. The survey includes papers from various journals and discusses recent trends and competitions in the field, as well as future directions for research.

Keywords: Natural Languages Processing (NLP) · Word Sense Disambiguation (WSD) · Approaches of WSD

1 Introduction

Human language is inherently ambiguous. In both natural languages and computational linguistics, it is quite common to encounter semantic ambiguity, where a single word can have multiple meanings. This phenomenon is pervasive across languages, and many words can be associated with more than one definition. For instance, let's take an example here:-

Consider the word "plant" in the following two sentences:

1. I watered the plant on my windowsill this morning.
2. There is a large textile plant in the outskirts.

In the first statement, "plant" refers to a living organism of the kind exemplified by trees that grows in soil and requires water to thrive, whereas "plant" refers to a vast industrial facility where goods are created, used in the second sentence. WSD would be utilised in each phrase to disambiguate the sense of "plant" by assessing the surrounding context and finding the most appropriate meaning for the word based on that context.

© The Author(s), under exclusive license to Springer Nature Switzerland AG 2024
S. Satheeskumaran et al. (Eds.): ICICSD 2023, CCIS 2122, pp. 55–70, 2024.
https://doi.org/10.1007/978-3-031-61298-5_5

Thus for a given word, there can be more than one sense and are called polysemous words while words with a single sense are called monosemous words. While humans can easily identify the intended meaning of a word in a given context, the same task is considerably challenging for a machine to accomplish. WSD and lexicography are closely related fields that can complement each other to enhance the quality of dictionaries and lexical resources. Lexicographers often write dictionary definitions that aim to capture various word senses. WSD can help lexicographers choose the most appropriate sense for each context. By analyzing a corpus of text, WSD algorithms can suggest which senses are commonly used in different contexts, allowing lexicographers to refine and improve the definitions.

In 1949, [38] Weaver introduced the problem of WSD in machine translation in his famous Memorandum on Machine Translation(MT). He emphasised that looking at the context of a term can help resolve the issue of multiple interpretations. Despite the development of statistical and machine learning methods for NLP, word polysemy remains a challenging and ubiquitous linguistic phenomenon.

This paper is divided into five sections. The Sect. 1 discusses the applications of WSD, while the Sect. 2 summarises recent papers and related work. The Sect. 3 outlines various approaches and the models used and datasets which are commonly used for WSD. Section 4 presents a tabular representation of the survey according to the datasets, and Sect. 5 concludes with potential future research directions. WSD is used in practically all language technology applications as seen in Fig. 1. Following are a few of the most popular applications of WSD.

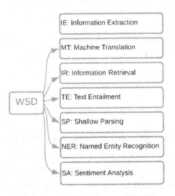

Fig. 1. Applications of WSD

Information Extraction (IE) aims to extract structured and useful data from sources that lack organization or have partial structure. WSD helps in correctly identifying the intended sense of any word in unstructured data. This is especially important for queries with ambiguous terms. Moreover, in question-answering systems, understanding the question's intent and the context of the provided text is crucial. It also helps in refining sentiment analysis, precise relation extraction and improving semantic understanding.

Machine Translation (MT) relies on WSD because certain words in every language can have varying translations depending on their contextual usage. WSD makes lexical choices for terms that have alternative interpretations for various senses. Source language (English): "I saw a bat flying in the sky." Target language (Spanish): "Vi un murciélago volando en el cielo." In this case, the word "bat" has numerous meanings in English, thus the machine translation system must determine the correct meaning based on the context. It might allude to a wooden bat or bat used in cricket or baseball, or it might allude to a flying animal. The machine translation system should accurately translate "bat" to "murciélago" in Spanish, which means "bat" in the sense of the animal, assuming the correct sense in this context is the flying mammal.

Sentiment Analysis (SA) is a process that makes use of computer algorithms to automatically recognize and extract the feelings and views represented in a text or voice. It aids in figuring out if the text's general attitude is neutral, positive or negative.

The strategy for WSD in Word-Class based selectional preferences that Xuri Tang et al. described in [45] uses association computation and the minimum description length for the WSD task while taking advantage of syntagmatic and paradigmatic semantic redundancy in the semantic system.

2   Related Work

There are numerous studies on WSD, some of which provide adequate accuracy for various languages but there hasn't been much analysis of these approaches. In this section, we've discussed the related work done till now.

The work proposed by Agirre et al. [2] combined a set of techniques which could be further extended to perform noun sense disambiguation. They used several unsupervised techniques that drew knowledge from a variety of sources.

Banerjee et al. [3] introduced a modified version of Lesk's dictionary-based word sense algorithm that utilized WordNet's lexical database instead of a regular dictionary. This enabled the algorithm to access a wide range of semantic relations present in the hierarchy of WordNet. In a similar vein, Montoyo et al. [4] combined the two primary approaches of knowledge-based and corpus-based methods to improve disambiguation results. The focus of their research was on the integration of these approaches and different sources of information to achieve better accuracy in word sense disambiguation. Martinez et al. [5] discovered that some algorithms performed well for some kind of words while each of them failed for others, upon studying and analyzing the relative results on customary algorithms in the same dataset.

Mihalcea et al. [6] demonstrated that the sense annotations obtained from Wikipedia are reliable and effective in developing accurate sense classifiers through Word Sense Disambiguation (WSD) experiments. Khapra et al. [7] proposed that a language with abundant resources could assist a less resourced language, they proposed a method to project corpus and WordNet parameters from a resourced language to a less resourced one, even when sense-tagged corpora are not available in the target language. Sun et al. [8] commenced a new Word Sense Disambiguation (WSD) approach that utilizes association guidelines for building an association rules-based database. The use of Genetic algorithm to figure out the appropriate meanings of polysemous nouns in the given context for Hindi Language was first explored by Sabnam et al. [12]. Their major goal was to use Genetic Algorithm to eliminate ambiguity in the sense using context.

Gutiérrez et al. [13] proposed a new unsupervised method that aims to automatically determine the intended meaning of a word in a specific context for different languages. Bala and team [14] discovered that combining a knowledge-based approach with selectional restrictions can prevent the formation of component word meaning representations that include selectional restrictions. They applied this to create a WSD tool for Hindi WordNet.

In Semantic Role Classification (SRC) research, the issue of ambiguity resulting from the limited impact and sparsity of lexical features is a recognized open problem, according to the proposal made by Benat Zapirain et al. [15]. To address this challenge, they employed models that integrate selectional preferences acquired through automated means. The researchers demonstrated that the Semantic Role Classification task is more effectively modeled by SP models that focus on both verbs and prepositions rather than on verbs alone, which helps to mitigate the ambiguity issue. Udaya Raj Dhungana et al. [17] introduced a new WordNet model for determining a polysemy word's correct sense based on the clue words. In their research, Sallam Abualhaija et al. [20] suggested two versions of the D-Bees algorithm to integrate domain knowledge into the disambiguation process. Devendra Singh Chaplot et al. [21] formalized the subject model to create a Word Sense Disambiguation (WSD) system that scales linearly with the number of words in the context.

In order to improve generalisation in scenarios with limited data, semantic memory captures past experiences seen during the model's lifespan. In the work of Zobaed et al. [26], they modelled the semantic link between the target word and the collection of glosses using both the context and associated gloss information of the target word. To complete the WSD job, they suggested SensPick, a kind of stacked bidirectional Long Short Term Memory (LSTM) network. Looking upon on the statistical relationship between the distribution of word sense numbers and word frequency rank Su et al. [28] provided a Z-reweighting strategy at the word level to alter the training on the unbalanced dataset based on the relation.

George et al. [34] introduced a new unsupervised WSD algorithm that makes use of all different kinds of semantic relationships found in word thesauruses. The algorithm uses Spreading Activation Networks (SANs), but unlike earlier WSD work, it uses a novel edge-weighting technique and builds SANs that take into account all sense-to-sense linkages rather than only relations between sensations and glosses. Marco et al. [33] analyze the limitations of current evaluation benchmarks in capturing the true capabilities of state-of-the-art WSD systems.

According to a study by Ahmed H. Aliwy et al. [41], which featured a comparison of different algorithms and techniques on WSD, concluded that no comparison can be considered accurate because every method was used on a separate dataset with a different size and Eloquence phenomena in some languages, such as the Arabic language, have an impact on how well the algorithm works.

Some recent articles have suggested the use of transfer learning [11], which can enhance the functionality of WSD systems by converting previously trained models to new domains or languages.

In their paper, Campolungo et al. [32] have discussed about DiBiMT, which they call the first entirely manually curated evaluation benchmark. It allows for a thorough investigation of semantic biases in machine translation of nominal and verbal words in five different language pairings, including English and one or more like Chinese, German, Italian, Russian or Spanish. Pal et al. in [40] assessed the performance's state of the art, recent works in several Indian languages, and conducted a survey in Bengali.

In [42], Sharma et al. created a WSD tool by utilising WordNet of Hindi and a knowledge-based method, LESK algorithm.

In order to create semi-supervised WSD systems, Osman et al. [46] suggested a fully-unsupervised method that automatically learns the many meanings of a word based on how it is used, with a limited quantity of sense-annotated data.

Flow of WSD Techniques

Figure 2 shows the flow of the WSD Techniques. At first, input with ambiguous words is taken. Then preprocessing is performed where the stop words, stem or lemmatize words are removed.

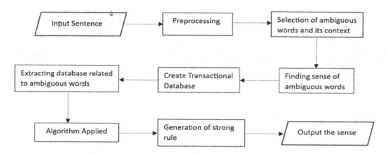

Fig. 2. Flow of WSD Techniques

3 Main Approaches and Datasets Used for WSD

The information source employed in word disambiguation is used to classify methodologies and strategies for WSD. Figure 3 depicts the primary techniques and associated algorithms.

3.1 Dictionary-Based or Knowledge-Based Approach

The knowledge-based method, a crucial element of WSD, makes use of external lexical resources. Knowledge-based approaches draw on a variety of knowledge sources, including thesauri, machine-readable dictionaries and sense inventories. Below, we have described the Knowledge based algorithms used till now in the survey papers.

Lesk Algorithm. The LESK algorithm utilizes definition overlap to simultaneously determine the appropriate senses for all words in a given context. Banerjee et al. [3]

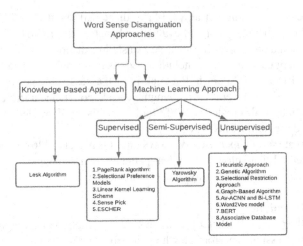

Fig. 3. Approaches used in WSD

proposed a modified version of the Lesk Dictionary Based WSD algorithm that employs WordNet's lexical database. This modification allows the algorithm to utilize the deep hierarchy of semantic relationships within WordNet, providing it with a powerful tool for semantic analysis.

3.2 Machine Learning Approach

Machine learning methods use the target word and context of its placement in the text as features. Unviewed instances are given senses and traits by a classifier to learn traits. Machine learning methods can be divided into unsupervised and supervised methods.

Supervised Approach

The supervised method uses training corpora that have been sense-annotated to discern between words' senses. The most successful supervised learning algorithms for WSD are support vector machines and memory-based learning. Some of the typical supervised learning algorithm types are listed below:

Iterative. WSD and PageRank Algorithm. The study in [7] makes it evident that IWSD is a greedy algorithm. It ignores words with a higher degree of polysemy and bases its conclusions on words that have already been clarified.

Linear Kernel Learning Scheme. The text is classified using the kernel support vector machine learning algorithm in the work proposed in [36], while feature extraction is handled via cuckoo search optimization. The movie dataset is treated to both favorable and negative reviews in order to increase classification accuracy. The IMDB dataset is used for the experiment.

ESCHER. In the paper [39], the ESCHER (Extractive Sense Comprehension) system is introduced as a supervised neural architecture for Word Sense Disambiguation (WSD).

This system employs a transformer-based approach that frames WSD as a span extraction task, where the goal is to identify the span that corresponds to the gloss of the correct sense for a given target word. To achieve this, ESCHER utilizes a pseudo-document created by concatenating the context of the target word with all of its potential senses.

Unsupervised Approach

Unsupervised WSD is an approach that uses unlabeled corpora and does not rely on pre-defined sense labels or categories. It is suitable for online information retrieval (IR) and machine translation (MT). The purpose of developing word occurrence clusters is to classify newly detected occurrences into the derived clusters. The algorithms listed below fall into the unsupervised category.

Heuristic Approach. The argument made in the paper [2] is that words can be disambiguated in a free-running text by combining the right set of heuristics. When a word is a component of a multi-word term, the heuristic H1 (Multi-words) is used. In this instance, the multi-word term's Hector sense is assigned. Just Hector's senses H1 through H8 are available. According to Heuristic H2 (Entry Sense Ordering), senses are arranged in an entry according to how frequently they are used.

Genetic Algorithm. The Genetic Algorithm optimizes the output of a separate keyword classifier as well as local context features (rather than also optimizing the keyword features together with the local context features). The usage of grammatical relation and chunk features is another improvement over previous versions of memory-based WSD. In the paper [13], a genetic algorithm is employed to determine the appropriate meanings of polysemous nouns in the given context. It is the first attempt to use a Genetic Algorithm for Hindi.

Graph-Based Algorithm. In the paper [16], a graph-based technique for automatic Word-net domain segmentation is presented in the study and demonstrates how the issue of ambiguity can be resolved using current graph-based image processing algorithms. In the paper [22], a graph convolution neural network (GCNN)-based system is trained to identify the domain of a given document using tokens and their PoS and dependency links as characteristics.

Bi-LSTM and Av-ACNN. In the paper [29], Bidirectional Long Short-Term Memory (Bi-LSTM) and Average Asymmetric Convolutional Neural Networks (Av-ACNN) are proposed as sequence processing models for improving Word Sense Disambiguation (WSD) accuracy in the biomedical field. The models utilize a combination of a convolutional neural network (CNN) and two LSTMs - one processing input in a forward direction and the other in a backward direction.

BERT. Paper [23] experimented with BERT-base, BERT-large, DistilBERT, Big Bird, RoBERTa, and ALBERT. In these experiments, the latent representation was extracted by selecting the transformer with the best-hidden layer. Best-hidden layer chosen for latent representation. Paper [27] proposes a BERT-based sequence labeling model with a transformer encoder above BERT to capture full sentence information.

Associative Database Model. A sophisticated recommendation system technique called collaborative filtering effectively modifies recommendations to match user preferences for associative database model. The proposed approach in [35] performed well on two datasets, their own created Movie-related dataset and Movielens dataset by IMDB which overcame all key problems to the recommendation system. It is important to include older-rated products when assuming user preferences.

Semi-supervised Algorithms

A form of machine learning algorithm known as semi-supervised learning lies between supervised and unsupervised learning algorithms. During the training phase, it uses a combination of labelled and unlabeled datasets where majority of the data is unlabeled. We found Yarowsky Algorithm which was used in the paper [11]. *Yarowsky Semi-supervised learning Algorithm.* Mishra et al. in [11] proposed that a semi-supervised approach combining stop word removal and stemming to the Yarowsky algorithm can provide a pre-eminent performance in the case of Hindi WSD. Dayu Yuan et al. [18] used a neural network language model to learn sequential and syntactic patterns from unlabeled text. The output of the language model is then used as features for a semi-supervised label propagation classifier. The label propagation classifier is trained on a small corpus of manually-disambiguated text which is then used to disambiguate words in a new text.

There have been numerous datasets used to test and evaluate WSD algorithms. Below are some examples of commonly used datasets that have also gained popularity since they can enhance WSD systems functionality by offering more details about word meanings.

Princeton WordNet is a vast, manually managed English lexicographic database that acts as the de facto benchmark inventory for WordSemantic Data. It is arranged as a network with synsets or collections of contextual synonyms, as nodes. Synsets and senses are connected to one another using edges that express lexical-semantic relations. The most current English WSD works use the 3.0 version, which was published in 2006 and has 117,659 synsets.

SenseEval is a set of evaluations focusing on detecting the correct sense of a word in context automatically. The assessments were organized as part of the SemEval event, providing researchers with a platform to compare their approaches and progress the state of the art in WSD. The SemCor corpus comprises 352 documents that are annotated with senses manually, producing a sum of 226,040 sense annotations.

4 Evaluation of the Survey

Survey results of supervised, unsupervised, knowledge-based and semi-supervised approaches can provide valuable insights into their performance on various word senses and situations. Supervised approaches may perform better on ambiguous words with clear contextual cues, while knowledge-based approaches may perform better on rare or domain-specific words with limited contextual information. Semi-supervised approaches may perform better on certain types of word senses, such as those related to named entities or specific domain terminology. Further research is needed to improve the performance of WSD models.

4.1 Supervised Approach

Below is the table for the survey of all the supervised approaches.

Name of the author [Year of publication] (Reference number)	Dataset used	Model Used	Accuracy
Andres Montoyo et al. [2005] [4]	Senseval-2 Spanish Data	Maximum Entropy Method	67.1% Precision
Mitesh M. Khapra et al. [2009] [7]	WordNet	iterative WSD (greedy), PageRank (exhaustive)	75% F1 score
Benat Zapirain et al. [2013] [15]	WordNet	Selectional Preference Models for Argument Classification	82.15% F1 score
Howard Chen et al. [2021] [24]	SemCor	non-parametric few-shot learning approach, Episodic Sampling algorithm	87.00% F1 score
Ying Su, Hongming Zhang et al. [2022] [28]	SemCor 3.0	supervised learning, Z-reweighting method and large pre-trained language models	78.6% F1 score
S. Zobaed et al. [2021] [26]	Utilized SemCor 3.0	Sense Pick	72.2% F1 score
P Ramya et al. [2023] [36]	SensEval, SensEval-, SenseEval-07, SenseEval-15	kernel support vector machine learning algorithm	93.8%, 92.5%, 94.5%, 94.5% F1 score respectively for the datasets
Rada Mihalcea [2007] [6]	WIKIPEDIA	Naive Bayes classifier	84.65% F1 score
Carmen Baneaa et al. [2011] [9]	English WordNet, Romanian WordNet	Multilingual subjectivity sense learning	69% macro accuracy
N. Mishra et al. [2012] [11]	Hindi Word Net	Yarowsky Algorithm	61.7 precision
Udaya Raj Dhungana et al. [2015] [17]	Nepali	WordNet	88.06% accuracy

4.2 Unsupervised Approach

We have listed the survey of the Unsupervised approaches below.

Name of the author [Year of publication] (Reference number)	Dataset used	Model Used	Accuracy
David Yarowsky [1995] [1]	English text	Iterative bootstrapping approach	96% F1 score
David Martinez, Eneko Agirre, Xinglong Wang [2006] [5]	Senseval 2, 3, Wordnet	Unsupervised learning, relatives in context	Based on scoring software provided by senseval 2, 42.4
Yong-le Sun, Ke-liang Jia [2009] [8]	ambiguous words from Center for Chinese Linguistics PKU and HowNet2008	Mining Association Rules	Precision score of 84.6% achieved
Yoan Gutirrez, Yenier Castaneda, Andy Gonzalez et al. [2013] [13]	BabelNet, WordNet	Personalised Page Rank combined with Frequencies of senses	69% of recall
Prity Bala [2013] [14]	Hindi wordnet	Selection restriction Algorithm	66.92% of accuracy
Brijesh Bhatt, Subhash Kunnath, Pushpak Bhattacharyya [2014] [16]	Wordnet	Graph Based Algorithm	F-Score of 0.74
Mikael Kågebäck, Hans Salomonsson [2016] [19]	SensEval	Bidirectional long short-term memory network	F1 score of 72.9
Devendra Singh Chaplot, Ruslan Salakhutdinov [2018] [21]	Senseval-3, Senseval-2, SemEval-2007, SemEval-2013 and SemEval-2015	unsupervised knowledge-based WSD	F1 score of 70.3 on SemEval-15
Alapan, Ayan Das and Sudeshna Sarkar [2020] [22]	English, Bengali, Hindi language	Un-Supervised Graph-based Approach	70% F1 score
Pawar, Siddhesh & Thombre et al. [2021] [23]	SemEval-2007 Task 6 dataset	Multi-Layer Perceptron classifier, BERT-base, BERT-large, DistilBERT, Big Bird, RoBERTa, and ALBERT	86.85% F1 score

(continued)

(*continued*)

Name of the author [Year of publication] (Reference number)	Dataset used	Model Used	Accuracy
Nithin Holla, Xiantong Zhen, Yingjun Du et al. [2021] [25]	SemCor corpus	GloVe embeddings, ELMo embeddings, BERT, β-VSM, variational semantic memory	78.8% F1 score
Jeong Yeon Park et al. [2022] [27]	English Senseval and Semeval datasets	Sequence labeling models, Sense Definition Clustering, BERT-based model	62.9%F1 score
Chun-Xiang Zhang, Shu-Yang Pang et al. [2022] [29]	MSH	Av-ACNN and Bi-LSTM	91.38%F1 score
Shreya Patankar, et al. [2022] [31]	SemEval 2010	word2Vec	48.9% recall
Md Samsuddoha, Dipto Biswas, Md. Erfan [2023] [35]	Own created Movie related dataset and a popular existing Movielens dataset developed by IMDB	Associative Database Model (ADBM) and the linguistics WSD approach	96.1% F1 score
Izunna Okpala, Guillermo Romera Rodriguez, Andrea Tapia, Shane Halse, Jess Kropczynski [2023] [37]	Stanford Contradiction Corpora dataset	Natural Language Processing (NLP) technique, lexicon-based approach, and sequence labelling	80% F1 score
E. Agirre et al. [2000] [2]	WordNet	Heuristics approach	98% of coverage
Sabnam Kumari et al. [2013] [12]	Hindi Language	Genetic Algorithm	91.60% of recall
Sallam Abualhaija et al. [2016] [20]	Chinese, Dutch, English, and Italian	D-bees algorithm	44.56% of precision
Surbhi Bhatia et al. [2022] [30]	Hindi WordNet	proposed genetic algorithm	80% F1 score

4.3 Knowledge-Based Approach

We have listed the survey of the knowledge-based approaches below.

Name of the author [Year of publication] (Reference number)	Dataset used	Model Used	Accuracy
Satanjeev Banerjee et al. [2002] [3]	WordNet, Senseval-2	Adapted Lesk Algorithm	Accuracy of 32%
Erwin Fernandez-Ordonez, et al. [2012] [10]	SEMEVAL 2007, English WordNet	Simplified Lesk algorithm	Accuracy of 64%
Nisheeth Joshi et al.[2019] [42]	Hindi WordNet	Lesk Algorithm	71.4% accuracy
Eneko Agirre et al. [2018] [43]	WordNet	UKB (Graph Based WSD)	67.3 F1 score
Eneko Agirre et al. [2009] [44]	WordNet	Personalised PageRank	51.5% accuracy

4.4 Semi-supervised Approach

We have listed the survey of the Semi-Supervised approaches below.

Name of the author [Year of publication] (Reference number)	Dataset used	Model Used	Accuracy
Julian Richardson et al. [2016] [18]	standard SemEval all-words tasks using WordNet as the inventory	LSTM algorithm	F1 score of 0.843
Xuri TANG et al. [2010] [45]	Chinese WordNet	Syntagmatic and Paradigmatic Redundancy Principle	69.88% accuracy
Osman Baskaya et al. [2016] [46]	WordNet	Ensembled WSID model	0.93 of F1 score
Duarte et al.[2021] [47]	Semeval-2007	ELECTRA	75.8% accuracy
Pratibha Rani et al. [2017] [48]	Hindi Tourism	Generic Semi Supervised approach	76.18 F1 score
Samuel Sousa et al. [2020] [49]	Senseval-2 LS, Senseval-3 LS, Semeval-2007	SSL Algorithms (word embeddings from Word2Vec, FastText, and BERT model)	87.5 F1 score
Kaveh Taghipour et al. [2015] [50]	SensEval-2 (SE2)	Support Vector Machines (SVM)	73.4% accuracy

5 Conclusion and Future Work

Although there have been significant advancements in the field of WSD, there are still several challenges that are notable. A serious issue would arise for different languages, where the structures and nuances of various languages may differ, necessitating the use of various methods and algorithms for WSD. Moreover, without domain-specific knowledge, disambiguating certain word meanings within different contexts can be challenging. Moreover, dealing with idiomatic expressions is a challenging task as the correct meaning of the expression is not evident from the individual words that make up the expression.

Through this review, we can infer that several new techniques and approaches have emerged in the field of WSD with each method used on a distinct data set with a different size. Models like BERT and other transformer-based architectures have shown remarkable performance in WSD. Specific adaptations of BERT and ELMo are finetuned on sense-annotated data and designed to handle the multi-sense nature of words. This is particularly valuable for low-resource languages. Better evaluation methods must be developed that are more robust and representative of real-world scenarios with balance accuracy with practical considerations such as scalability, domain specificity and computational efficiency.

References

1. Yarowsky, D: Unsupervised word sense disambiguation rivaling supervised methods. In: Proceedings of ACL 1995, pp 189–196, Cambridge, Massachusetts (1995)
2. Agirre, E., Rigau, G., Padro, L., Atserias, J.: Combining supervised and unsupervised lexical knowledge methods for word sense disambiguation. Comput. Humanit. **34**, 103–108 (2000)
3. Banerjee, S., Pedersen, T.: An adapted Lesk algorithm for word sense disambiguation using WordNet. In: Computational Linguistics and Intelligent Text Processing. LNCS, vol. 2276, pp. 136–145. Springer, Heidelberg (2002). https://doi.org/10.1007/3-540-45715-1_11
4. Montoyo, A., Rigau, G., Suárez, A., Palomar, M.: Combining knowledge- and corpus-based word-sense-disambiguation methods. J. Artif. Intell. Res. **23**, 299–330 (2005)
5. Martinez, D., Agirre, E., Wang, X.: Word relatives in context for word sense disambiguation. In: Proceedings of the 2006 Australasian Language Technology Workshop (ALTW2006), pp. 42–50 (2006)
6. Mihalcea, R.: Using Wikipedia for automatic word sense disambiguation. In: Human Language Technologies 2007: The Conference of the North American Chapter of the Association for Computational Linguistics; Proceedings of the Main Conference, pp. 196–203. Association for Computational Linguistics, Rochester (2007)
7. Khapra, M.M., Shah, S., Kedia, P., Bhattacharyya, P.: Projecting parameters for multilingual word sense disambiguation. In: Proceedings of the 2009 Conference on Empirical Methods in Natural Language Processing, pp. 459–467. Association for Computational Linguistics, Singapore (2009)
8. Sun, Y., Jia, K.-L.: Research of Word Sense Disambiguation Based on Mining Association Rule, pp. 86–88. Third International Symposium on Intelligent Information Technology Application Workshops, Nanchang (2009)
9. Banea, C., Mihalcea, R., Wiebe, J.: Sense-level subjectivity in a multilingual setting. In: Proceedings of the workshop on sentiment analysis where AI meets psychology (SAAIP 2011), pp. 44–50. Asian Federation of Natural Language Processing. Chiang Mai (2011)

10. Fernandez-Ordoñez, E., Mihalcea, R., Hassan, S.: Unsupervised word sense disambiguation with multilingual representations. In: Proceedings of the Eighth International Conference on Language Resources and Evaluation (LREC 2012), pp. 847–851. European Language Resources Association (ELRA), Istanbul (2012)

11. Mishra, N., Siddiqui, T.J.: An investigation to semi supervised approach for HINDI word sense disambiguation. In: Trends in Innovative Computing Intelligent Systems Design (2012)

12. Kumari, S., Singh, P.: Optimized word sense disambiguation in Hindi using genetic algorithm. Int. J. Res. Comput. Commun. Technol. 2(7), 445–449 (2013)

13. Gutiérrez, Y., et al.: UMCC_DLSI: Reinforcing a Ranking Algorithm with Sense Frequencies and Multidimensional Semantic Resources to solve Multilingual Word Sense Disambiguation (2013)

14. Bala, P.: Word sense disambiguation using selectional restriction. Int. J. Sci. Res. Publ. 3(4) (2013)

15. Zapirain, B., Agirre, E., Marquez, L., Surdeanu, M.: Selectional preferences for semantic role classification. Comput. Linguist. 39, 631–663 (2013)

16. Bhatt, B., Kunnath, S., Bhattacharyya, P.: Graph based algorithm for automatic domain segmentation of WordNet. In: Proceedings of the Seventh Global Wordnet Conference, pp. 178–185. University of Tartu Press, Tartu (2014)

17. Dhungana, U.R., Shakya, S., Baral, K., Sharma, B.: Word sense disambiguation using WSD specific WordNet of polysemy words. In: Proceedings of the 2015 IEEE 9th International Conference on Semantic Computing (IEEE ICSC 2015), Anaheim, 2015, pp. 148–152 (2015)

18. Yuan, D., Richardson, J., Doherty, R., Evans, C., Altendorf, E.: Semi-supervised word sense disambiguation with neural models. In: Proceedings of the 26th International Conference on Computational Linguistics: Technical Papers (COLING 2016), Osaka, pp. 1374–1385. The COLING 2016 Organizing Committee (2016)

19. Kågebäck, M., Salomonsson, H.: Word Sense Disambiguation Using a Bidirectional LSTM. CogALex@COLING (2016)

20. Abualhaija, S., Zimmermann, K.-H.: Solving specific domain word sense disambiguation using the D-Bees algorithm. Glob. J. Technol. Optimiz. (2016)

21. Chaplot, D.S., Salakhutdinov, R.: Knowledge-based word sense disambiguation using topic models. In: Proceedings of the AAAI Conference on Artificial Intelligence (2018)

22. Kuila, A., Das, A., Sarkar, S.: A graph convolution network-based system for technical domain identification. In: Proceedings of the 17th International Conference on Natural Language Processing (ICON): TechDOfication 2020 Shared Task (2020)

23. Pawar, S., Thombre, S., Mittal, A., Ponkiya, G., Bhattacharyya, P.: Tapping BERT for Preposition Sense Disambiguation (2021)

24. Chen, H., Xia, M., Chen, D.: Non-parametric few-shot learning for word sense disambiguation. In: Proceedings of the 2021 Conference of the North American Chapter of the Association for Computational Linguistics: Human Language Technologies (2021)

25. Du, Y., Holla, N., Zhen, X., Snoek, C.G.M., Shutova, E.: Meta-learning with variational semantic memory for word sense disambiguation. In: Annual Meeting of the Association for Computational Linguistic (2021)

26. Zobaed, S., Haque, M.E., Rabby, M.F., Salehi, M.A.: SensPick: sense picking for word sense disambiguation. In: Proceedings of 2021 IEEE 15th International Conference on Semantic Computing (ICSC) (2021)

27. Park, J.Y., Shin, H.J., Lee, J.S.: Word sense disambiguation using clustered sense labels. Appl. Sci. 12(4), 1857 (2022)

28. Su, Y., Zhang, H., Song, Y., Zhang, T.: Rare and zero-shot word sense disambiguation using Z-reweighting. In: Proceedings of the 60th Annual Meeting of the Association for Computational Linguistics, vol. 1: Long Papers, Dublin, pp. 4713–4723 (2022)

29. Zhang, C.-X., Pang, S.-Y., Gao, X.-Y., Jia-Qi, L., Bo, Y.: Attention neural network for biomedical word sense disambiguation. Discrete Dyn. Nat. Soc. Hindawi **2022**, 1–14 (2022)
30. Bhatia, S., Kumar, A., Khan, M.M.: Role of genetic algorithm in optimization of Hindi word sense disambiguation. IEEE Access **10**, 75693–75707 (2022)
31. Patankar, S., Phadke, M., Devane, S.: Wiki sense bag creation using multilingual word sense disambiguation. IAES Int. J. Artif. Intell. **11**(1), 319–326 (2022)
32. Campolungo, N., Martelli, F., Saina, F., Navigli, R.: DiBiMT: a novel benchmark for measuring word sense disambiguation biases in machine translation. In: Proceedings of the 60th Annual Meeting of the Association for Computational Linguistics (Volume 1: Long Papers), pp. 4331–4352 (2022)
33. Maru, M., Conia, S., Bevilacqua, M., Navigli, R.: Nibbling at the hard core of word sense disambiguation. In: Proceedings of the 60th Annual Meeting of the Association for Computational Linguistics (Volume 1: Long Papers), pp. 4724–4737 (2022)
34. George, T., Vazirgiannis, M., Androutsopoulos, I.: Word sense disambiguation with spreading activation networks generated from Thesauri. In: IJCAI International Joint Conference on Artificial Intelligence, pp. 1725–1730 (2007)
35. Samsuddoha, M., Biswas, D., Erfan, M.: User Similarity Computation Strategy for Collaborative Filtering Using Word Sense Disambiguation Technique, pp. 87–101. Springer, Singapore (2023). https://doi.org/10.1007/978-981-19-8032-9_7
36. Ramya, P., Karthik, B.: Word sense disambiguation based sentiment classification using linear kernel learning scheme. Intell. Automat. Soft Comput. (2022)
37. Okpala, I., Rodriguez, G.R., Tapia, A., Halse, S., Kropczynski, J.: A semantic approach to negation detection and word disambiguation with natural language processing. In: Proceedings of the 6th International Conference on Natural Language Processing and Information Retrieval (NLPIR 2022) (2022)
38. Weaver, W.: MT News International, no. 22, July 1999, pp. 5–6, 15 (1949)
39. Edoardo, B., Pasini, T., Navigli, R.: ESC: redesigning WSD with extractive sense comprehension. In: Proceedings of the 2021 Conference of the North American Chapter of the Association for Computational Linguistics: Human Language Technologies, pp. 4661–4672 (2021)
40. Pal, A.R., Saha, D.: Word sense disambiguation: a survey. Int. J. Control Theory Comput. Model. **5**(3), 1–6 (2015)
41. Aliwy, A.H., Taher, H.A.: Word sense disambiguation: survey study. J. Comput. Sci. **15**(7), 1004–1011 (2019)
42. Sharma, P., Joshi, N.: Knowledge-based method for word sense disambiguation by using Hindi WordNet. Eng. Technol. Appl. Sci. Res. **9**(2), 3985–3989 (2019)
43. Agirre, E., de Lacalle, O.L., Soroa, A.: The risk of sub-optimal use of open source NLP software: UKB is inadvertently state-of-the-art in knowledge-based WSD. In: Proceedings of Workshop for NLP Open Source Software, Melbourne, 20 July 2018, pp. 29–33 (2018)
44. Agirre, E., de Lacalle, O.L., Soroa, A.: Knowledge-based WSD on specific domains: performing better than generic supervised WSD. In: Proceedings of the 21st International Joint Conference on Artificial Intelligence (IJCAI'09), pp. 1501–1506 (2009)
45. Tang, X., Chen, X., Qu, W., Yu, S.: Semi-supervised WSD in selectional preferences with semantic redundancy. Coling 2010: Poster Volume, pp. 1238–1246 (2010)
46. Başkaya, O., Jurgens, D.: Semi-supervised learning with induced word senses for state of the art word sense disambiguation. J. Artif. Intell. Res. **55**, 1025–1058 (2016)
47. Duarte, J.M., Sousa, S., Milios, E., Berton, L: Deep analysis of word sense disambiguation via semi-supervised learning and neural word representations. Inf. Sci. **570**, 278–297 (2021). https://doi.org/10.1016/j.ins.2021.04.006

48. Rani, P., Pudi, V., Sharma, D.M.: Semisupervised data driven word sense disambiguation for resource-poor languages. In: 4th International Conference on Natural Language Processing (ICON 2017) (2017)
49. Sousa, S., Milios, E., Berton, L.: Word sense disambiguation: an evaluation study of semi-supervised approaches with word embeddings. In: 2020 International Joint Conference on Neural Networks (IJCNN), Glasgow, pp. 1–8 (2020). https://doi.org/10.1109/IJCNN48605.2020.9207225
50. Taghipour, K., Ng, H.: Semi-Supervised Word Sense Disambiguation Using Word Embeddings in General and Specific Domains, pp. 314–323 (2015). https://doi.org/10.3115/v1/N15-1035

Oral Cancer Classification Using GLRLM Combined with Fuzzy Cognitive Map and Support Vector Machines from Dental Radiograph Images

K. Anuradha[1]([✉]) [ID], H. Fathima[2], K. Kavithamani[2], and K. P. Uma[2]

[1] School of Computer Applications, Karpagam College of Engineering, Othakkal Mandapam, Tamil Nadu, India
k_anur@yahoo.com
[2] Department of Science and Humanities, Hindusthan College of Engineering and Technology, Coimbatore, Tamil Nadu, India

Abstract. In recent years, Fuzzy Cognitive Maps (FCM) are widely used in medical decision schemes. According to data, oral cancer is the fifth most common cancer in India, with a higher occurrence among men than women. In order to save lives, correct diagnosis and prompt treatment is required. When oral cancer is identified early, the 5-year survival rate exceeds 80%, but advanced stages of the illness have 5-year survival rates that are less than 20–30%. But, the majority of cases are detected in advanced stages where treatment becomes unsuccessful. To achieve a better classification of oral cancers, Gray Level Run Length Matrix (GLRLM) is combined with Fuzzy Cognitive Map and Support Vector Machines. In this work, fifty Dental Radiograph images are used for cancer classification. Initially, Linear contrast stretching algorithm is employed to remove the noise present in the image. The filtered picture is split to create a Region of Interest (ROI). The ROI picture is used to extract the GLRLM features. The classification of cancer is based on the features extracted. From the extracted features, a fuzzy cognitive map is employed. Cancer categorization is carried through Machine Learning approaches. Measures like sensitivity, specificity, precision, accuracy, and F-score have been used to describe the performance of the SVM classifier. The study's findings demonstrate that GLRLM-based features are strong discriminating features for statistical analysis of Dental Radiograph images and can be helpful in the identification of cancer. By combining the texture features based on GLRLM and FCM, the accuracy (94%) as a whole is improved.

Keywords: Fuzzy Cognitive Map · Gray Level Run Length Matric · Gray Level Co-occurrence Matrix · Radiograph Images · Support Vector Machine · Machine Learning · Image Processing

© The Author(s), under exclusive license to Springer Nature Switzerland AG 2024
S. Satheeskumaran et al. (Eds.): ICICSD 2023, CCIS 2122, pp. 71–81, 2024.
https://doi.org/10.1007/978-3-031-61298-5_6

1 Introduction

Oral cancer is the abnormal growth of cells present in the oral cavity. These abnormal cells replicate rapidly inside the mouth and affect other parts of the body [1]. Due to recent advancements in Medical field, cancer diagnosis has become easier. Reducing the delay in diagnosis is necessary to meet the goal of improving the prognosis for oral cancer because it is the primary contributor to a cancer diagnosis at a late stage. Oral cancer detection delays are caused by a number of complex factors, including patients, medical professionals, and health systems [2]. Oral cancers can be diagnosed using MRI, CT, Endoscopy, Biopsy, Barium swallow, Ultrasound, Positron emission tomography. Besides these techniques, a thorough physical examination is also necessary to invade the presence of oral cancer in mouth, neck and throat regions. Researchers used various staging techniques to classify cancers [3]. TNM classification method is the most common method used to stage cancers. The TNM staging system has been widely used for cancer staging and prognostication. However, limited flexibility in adapting to the evolving understanding of cancer biology and incorporating new prognostic variables [4].

The aim of this study is to diagnose oral cancers at an earlier stage. Also the study aims to explore the combination of Feature Extraction and Machine Learning Techniques for classification. Fifty Dental Radiograph images are taken for this research to classify it as Benign and Malignant. The database consists of Oral Squamous Oral Leukoplakia, Erythroplakia, Fibroma, Cystic carcinoma, Mucosal melanoma and Lichen Planus. Among these types, oral squamous cell carcinoma found to be very harmful [5]. This paper uses the GLRLM features to construct a Fuzzy Cognitive Map. The Long Run Emphasis (LRE), Gray Level Non-uniformity (GLN), are used for classification. The map relates these features (concepts) and once the value of each concept is calculated, it is important to analyze and interpret the results. It also uses value between the inter-relation between the concepts. Then the weights are adjusted using Active Hebbian Learning Algorithm. Finally, Machine Learning algorithms are used for classification of tumors.

This paper is structured as shadows: Related works are discussed in Sect. 2. The methodology used in this paper is discussed in Sect. 3. Results and Discussion are shown in Sect. 4. The paper is decided in Sect. 5.

2 Related Works

Fuzzy Cognitive Maps are widely used in Medical Fields [6]. Recently [6], authors used Fuzzy Cognitive Map to identify the main factor causing Covid-19. The symptoms causing the disease are identified initially. And from the symptoms, the root cause of the disease was identified. Authors in [7] made a complete review of Fuzzy Cognitive Maps in medicine. They investigated the different FCM models used in Medical decision support systems. They also reviewed various diagnosis systems.

Authors in [8] combined Fuzzy Cognitive Map with Support Vector Machines for the classification of Bladder Tumors. The histopathological features were used to construct FCM. In order to train the map, an unsupervised learning algorithm was used. This study achieved low grade accuracy of 85.54% and high grade accuracy of 89.13%. In comparison to conventional models, the proposed model was evaluated.

Previously, Anuradha et al., [9] used a combination of FCM and GLCM for the classification of oral cancers from Dental Radiographs. Five GLCM features were used to construct the FCM. Performance evaluation measures such as Accuracy, Sensitivity and Specificity were calculated. The calculated values were compared with methods not using FCM. Fuzzy Cognitive Maps were used to classify breast cancers [10–15]. The authors used Fuzzy Cognitive Maps combined with various techniques to classify breast cancers.

GLRLM feature extraction techniques are widely used in medical decision systems. Shilpa Harnale and Dhananjay Maktedar [16] used Histogram, GLCM and GLRLM feature extraction to classify oral neoplasms from MRI images. Finally, the three techniques were compared to find an efficient algorithm. Support Vector Machine was used for classification. Among these three algorithms GLRLM combined with SVM achieved an excellent result.

In order to classify oral cancer from dental radiograph images, Anuradha K and Sankaranarayanan K [17] examined feature extraction techniques (Intensity histogram, GLCM, and GLRLM). The authors used Supervised Learning Algorithm for classification accuracy. They finally concluded that the grouping of GLCM and SVM achieved a good accuracy than the other methods.

Anuradha et al. [24] used Fisher's tumor grading system to construct the model. Eight features are extracted from the grading systems. The authors achieved an accuracy of 94.11% for Low grade tumors and 92.10% for High grade tumors.

Like Anuradha K and Sankaranarayanan K [17], Durgamahanthi et al. [27] used GLCM and GLRLM features for the identification of brain cancer. A combination of GLCM and GLRLM features were also proposed by them to achieve a good accuracy. Histopathology images were taken for their study and the authors suggested that the information and result obtained from these techniques can be decided to perform brain surgery.

From the literature it is clear that the accuracy of cancer classification is obtained when pathological methods are used. But biopsy becomes harmful [25]. It has not yet been possible to classify OPMDs and OML using algorithms based on VELscope image analysis.

Previous studies have demonstrated the ability to distinguish between cancer and precancerous cells using a more quantitative study of the intensity pictures. There isn't a computer-based textural analysis approach to assist doctors in more quickly and accurately identifying ROIs [26].

3   Methodology

The workflow of the proposed methodology is shown in Fig. 1.

The Digitized X Ray images are taken as input. As the X-Ray images contains more noise, preprocessing is done to remove the image. Then segmentation is performed to clearly detect the tumor part from the image. From the segmentation image, features are extracted using Gray Level Run Length Matrix. Then these values are normalized and given as input to Fuzzy Cognitive Map for prediction and classification. The categories obtained are the intermediate values. Support Vector Machine classifier is used to classify

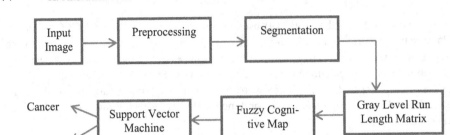

Fig. 1. Workflow of the Projected Model

the images as cancer or non-cancer. This study uses fifty digitized Dental Radiographs. The data set contains: Oral Squamous Cell Carcinoma (5), Oral Sub Mucous Fibrosis (7), Oral Leukoplakia (2), Erythroplakia (5), Fibroma (6), Cystic carcinoma (3), Mucosal melanoma (9) and Lichen Planus (13). Linear contrast stretching is used as the first preprocessing step for the input images. After improvement, the separately using the watershed transform. The features are then extracted using GLRLM. After modeling a fuzzy cognitive map, determining the tumor's condition, SVM is used to determine if it is cancerous or benign.

3.1 Image Preprocessing

Dental Radiographs contains more noise which becomes difficult to extract. Adding linear contrast to an image enhances it.

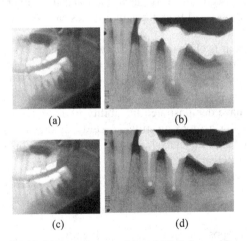

Fig. 2. (a) (b) Input Images. (c) and (d) Enhanced Images

The original digital values are gray scaled as it linearly expands, causing the tumor area to seem darker and the teeth and bone areas to appear lighter in the image. The tumour becomes more apparent as a result of this. All of the pictures are linearly contrasted here [17].

The Fig. 2(a) and Fig. 2(b) are the input images besides Fig. 2(c) and Fig. 2(d) are the preprocessed images.

3.2 Image Segmentation

The method of segmenting an image into several similar sections is called segmentation. Following pre-processing, the corrected images are partitioned using Watershed segmentation. The tumour will be segmented as a result of this. Watershed segmentation is a technique for extracting distinct image. Figure 2 shows the intermediate result of Watershed Transform for Fig. 2(c) and Fig. 2(d).

From Fig. 3, Region of Interest is segmented where the GLRLM features are extracted.

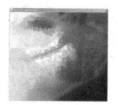 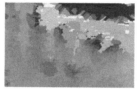

Fig. 3. Watershed Transform

3.3 Gray Level Run Length Matrix

Gray Level Run Length Matrix (GLRLM) is a method of extracting texture characteristics with higher properties. A grey level run is a line of pixels that moves in one direction. The length of the grey level run and the number of such pixels are determined by the number of such pixels. Few authors used GLRLM for oral cancer classification [18, 19, 23]. In this work, seven GLRLM features are extracted from the region of interest (ROI) image that has been segmented, together with the following parameters: run percentage, low gray level run emphasis (LGLRE), high gray level run emphasis (LGLRE), length of run non-uniformity (RLN), and gray level non-uniformity (GLN).

In [23], back propagation Artificial Neural Network and GLRLM along with Gray Level Co-occurrence Matrix and Intensity Histogram techniques used to (total 61 features) improve the classification accuracy.

3.4 Fuzzy Cognitive Map

Fuzzy Cognitive Maps are used extensively in Medical Fields. It was first developed by Kosko in 1986 [20]. It is possible to understand a fuzzy as a graphical representation of information hybrid of fuzzy logic and cognitive mapping. Cognitive mapping, which is also the basis of most calculations and indices, is based on graph theory. A FCM consists of factors that represent the essential elements (concepts/nodes). Experts knowledge is very essential to relate the concepts. Graph nodes are definitions that refer to the variables, states, factors and other features used in the perfect (that describe the behavior of the system). As shown in Fig. 4, causal relationships interconnect the FCM concepts. Figure 3 is a sample diagram for modeling the FCM.

In Fig. 4, C1, C2 to C9 are the concepts/features of Oral tumor [21]. The arrow mark in the diagram represents the relationship between the nodes.

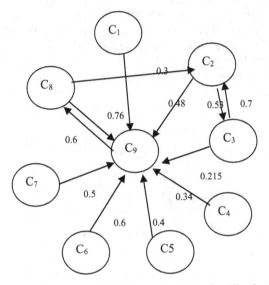

Fig. 4. Fuzzy Cognitive Map for Oral Tumor Grading using Histological Features

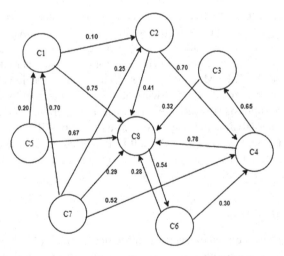

Fig. 5. FCM for oral tumor grading using GLRLM features

The node C2 is related to C3. A slight change in C2 will also deviate in C3. The negative value decreases the value.

Based on the experts opinion, the Fuzzy Cognitive Map is modelled using the GLRLM (Fig. 5). The features removed from the images are the concepts in FCM.

The features are related as follows:

C1 – Short Run Emphasis

C2 – Long Run Emphasis

C3 – Gray Level Non uniformity

C4 – Run Length non-uniformity

C5 – Run percentage

C6 – Low Gray Level Run emphasis

C7 – High Grapy Level Run Emphasis

The model (Fig. 5) consists of seven concepts and sixteen weight relations. Concepts C1–C7 are the GLRLM features. C8 is the decision node of oral tumor grading. After the FCM model has been developed and the necessary specifications for the implementation of the Active Hebbian Learning (AHL) algorithm have been determined, the FCM tumor-grading model can be used. AHL specifies the weight of the connection between the two concepts should be adjusted (increase/decrease) in proportion to their activation product. A less human intervention is required to determine the initial weights of the FCM. The value of Ai values of concepts connected to it at each step and is updated. The values are normalized and in the ranges 0–1. According to the model, the concept C4 (Run Length Non-uniformity) with weight 0.78 induces more of the concept C8.

Anuradha et al. [21] used histological Feature extraction to model FCM. An accuracy of 90.58% was achieved. This was achieved due to the Histological features used.

The model developed worked better with Histological features when compared to Image Extraction Features.

3.5 Support Vector Machine

The final step of the proposed work flow (Fig. 1) is classification. This work uses SVM for classification of tumors. This binary classifier classifies the category as "cancer" or "non cancer". Out of 50 images, 35 images are used for training the model. Here all the trained support vectors are perfectly linearly separable (two classes with margin). The python library sklearn is used to split the data set randomly. The data points (support

Table 1. Performance Measures

Measure	Derivations
Sensitivity	TPR = TP/(TP + FN)
Specificity	SPC = TN/(FP + TN)
Accuracy	ACC = (TP + TN)/(P + N)

Where TP is the True Positive (cancer is predicted as "yes")

TN is the True Negative (non-cancer is predicted as "no")

FP is the False Positive (non-cancer is predicted as "yes")

FN is the False Negative (cancer is predicted as "no")

The performance of any model can be evaluated using the above measures.

78   K. Anuradha et al.

vectors) are correctly spread identifying the cancer classes and benign classes. The premalignant lesions of Leukoplakia lies in the margin level making the model hard to classify. The support vectors near the margin are the pure Leukoplakia images. Thus, those data are considered as cancer data. The presentation of the model is calculated using the following measures:

4 Results and Discussions

Few images were obtained from the Section of Radiology at the Mid-Michigan Medical Center in Michigan [21]. There are 50 images (24 cancer and 26 non cancer) in the database. For these images, initially Linear contrast stretching was used for image enhancement. Seven GLRLM features were extracted from the images. Using these features, a model in constructed with the expert's knowledge. According to the expert's C8 is assigned a value greater than 0.80 for malignant cases and less than 0.80 for benign cases.

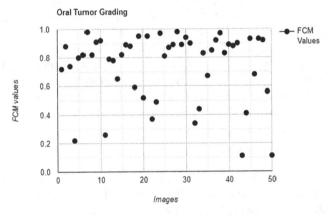

Fig. 6. FCM grade values for oral tumor

The SVM classifier classified 23 benign cases out of 26 and 24 cancer cases out of 24 from GLRLM features. Figure 6 shows the classification points where the support vectors greater than 0.8 are cancer cases and less than 0.8 are benign cases. The confusion matrix is exposed in Table 2:

Table 2. Confusion Matrix

Predicted	Actual		
		Cancer	Non-cancer
	Cancer	24 (TP)	2 (FP)
	Non-cancer	1 (FN)	23 (TN)

The performance measures (Accuracy, Sensitivity and Specificity) obtained are shown in Table 3 [16] which is calculated using Table 1. The SVM classifier is the best fit for all the cancer classes. Both in training and testing stages, the cancer cases are classified correctly with 100% accuracy. With only few samples (50 images) and 35 images are used for training itself, the model cannot classify correctly for benign cases. The model is over fitted as more training is provided with benign cases.

Table 3 demonstrates that the combination of other techniques outperformed FCM in terms of precision. The reason for this is the advancement in the watershed segmentation algorithm. FCM has an accuracy of 94 percent in this work. Though only five features were extracted, GLCM obtained a good accuracy with 96%. But from Table 2, it is observed that the perfect has classified cancer cases correctly. 24 cases (True Positives) has been predicted and classified (94%). But the pre-malignant or benign cases have not predicted correctly. Thus the study has limitations. Firstly, the no. of sample images is 50. With 50 images, the model has classified cancer cases correctly at an earlier diagnosis. Introduction of FCM in the work has improved an accuracy (94%).

Table 3. Presentation Measures

Technique	Accuracy	Specificity	Sensitivity
GLCM + FCM + SVM [22]	92%	91%	88%
GLRLM + FCM + SVM (this work)	94%	92%	96%
GLCM + SVM [17]	96%	100%	93%
GLRLM + SVM [17]	92%	95.45%	88%
IH + SVM [17]	88%	90%	85%

5   Conclusion

Advancements in medical imaging technology have opened up new possibilities for the early discovery and classification of diseases. Oral cancer is a severe and potentially fatal illness that impacts numerous individuals across the globe. Early detection is crucial for effective treatment and improving patient outcomes. Here, this work used a combination of GLRLM, FCM and SVM for oral cancer classification. GLRLM features were extracted from the dental X-Ray images containing benign and malignant cases. Here combining the above techniques, an accuracy of 94% was achieved. Improving the watershed segmentation will yield a good accuracy. A comparison of various methods was also made. In future, the sum of features and the sum of input images can be raised.

References

1. Abhishek, D.: Squemous cell carcinoma- uncontrolled growth of abnormal cells. J. Med. Sci. Clin. Res. **7**(7) (2019)
2. González-Moles, M.Á., Aguilar-Ruiz, M., Ramos-García, P.: Challenges in the early diagnosis of oral cancer, evidence gaps and strategies for improvement: a scoping review of systematic reviews. Cancers **14**(19), 4967 (2012)
3. Almangush, A., et al.: Staging and Grading of Oral squamous cell carcinoma: an update. Oral Oncol. **107** (2020)
4. Hubert Low, T.H., Gao, K., Elliott, M., Clark, J.R.: Tumor classification for early oral cancer: Re-evaluate the current TNM classification. Head Neck **37**(2), 223–228 (2014)
5. Jiang, X., Wu, J., Wang, J., Huang, R.: Tobacco and oral squamous cell carcinoma: a review of carcinogenic pathways. Tob. Induc. Dis. **17**, 29 (2019)
6. Kanchana, A., Varuvel, V.N.: A study of COVID-19 symptoms using fuzzy cognitive map. Int. J. Pharmaceut. Res. **4**(12), 1068–1072 (2020)
7. Amirkhani, A., Papageorgiou, E., Mohseni, A., Mosavi, M.R.: A review of fuzzy cognitive maps in medicine: taxonomy, methods, and applications. Comput. Methods Prog. Biomed. **142**, 129–145 (2017)
8. Papageorgiou, E., Georgoulas, G., Stylios, C., Nikiforidis, G., Groumpos, P.: Combining fuzzy cognitive maps with support vector machines for bladder tumor grading. In: Gabrys, B., Howlett, R.J., Jain, L.C. (eds.) KES 2006. LNCS (LNAI), vol. 4251, pp. 515–523. Springer, Heidelberg (2006). https://doi.org/10.1007/11892960_63
9. Anuradha, K.: Efficient oral cancer classification using GLCM feature extraction and fuzzy cognitive map from dental radiographs. Int. J. Pure Appl. Math. **118**(20), 651–656 (2018)
10. Amirkhani, A., Mosavi, M.R., Naimi, A.: Unsupervised fuzzy cognitive map in diagnosis of breast epithelial lesions. In: 22nd Iranian Conference on Biomedical Engineering (ICBME), Tehran, 2015, pp. 115–119 (2015)
11. Thuthi Sarabai, D., Arthi, K.: Efficient breast cancer classification using improved fuzzy cognitive maps with Csonn. Int. J. Appl. Eng. Res. **11**(4), 2478–2485 (2016)
12. Roopa Chandrika, R., Karthikeyan, N., Karthik, S.: Texture classification using fuzzy cognitive maps for grading breast tumor. Asian J. Inf. Technol. **15**(5), 989–995 (2016)
13. Janet Sheeba, J., Victor Devadoss, A., Albert William, M.: An analysis of risk factors of breast cancer using interval weighted fuzzy cognitive maps. Int. J. Comput. Algorithm **3**(2), 161–166 (2014)
14. Jayashree, S, Akila, K., Papageorgiou, E., Papandrianos, N., Vasukie, A.: An integrated breast cancer risk assessment and management model based on fuzzy cognitive maps. Comput. Methods Prog. Biomed. **118**(3), 280 (2015)
15. Christodoulou, P., Christoforou, A., Andreou, A.S.: A hybrid prediction model integrating fuzzy cognitive maps with support vector machines. In: Proceedings of the 19th International Conference on Enterprise Information Systems (ICEIS 2017), vol. 1, pp. 554–564 (2017). ISBN: 978-989-758-247-9
16. Harnale, S., Maktedar, D.: Oral cancer detection: feature extraction & SVM classification. Int. J. Adv. Netw. Appl. **11**(3), 4294–4297 (2019)
17. Anuradha, K., Sankaranarayanan, K.: Comparison of feature extraction techniques to classify oral cancers using image processing. Int. J. Appl. Innov. Eng. Manag. **2**(6), 456–462 (2013)
18. Anuradha, K., Sankaranaryanan, K.: Comparison of feature extraction techniques to classify oral cancers using image processing. Int. J. Appl. Innov. Eng. Manag. **2**(6), 456–462 (2013)
19. Rahman, T.Y., Mahanta, L.B., Choudhury, H., Das, A.K., Sarma, J.D.: Study of morphological and textural features for classification of oral squamous cell carcinoma by traditional machine learning techniques. Cancer Rep. (Hoboken, NJ), **3**(6), e1293 (2020). https://doi.org/10.1002/cnr2.1293

20. Kosko, B.: Fuzzy cognitive maps. Int. J. Man Mach. Stud. **24**(1), 65–75 (1986)
21. Anuradha, K., Uma, K.P.: Histological grading of oral tumors using fuzzy cognitive map. Biomed. Pharmacol. J. **10**(4), 1695–1700 (2017)
22. http://calcnet.mth.cmich.edu/org/spss/Prj_cancer_data.htm
23. Thomas, B., Kumar, V., Saini, S.: Texture analysis based segmentation and classification of oral cancer lesions in color images using ANN. In: IEEE International Conference on Signal Processing, Computing and Control (ISPCC), Solan, pp. 1–5 (2013). https://doi.org/10.1109/ISPCC.2013.6663401
24. Anuradha, K., Uma, K.P.: Implementation of fuzzy cognitive map and support vector machine for the classification of oral cancers. EAI Endors. Trans. Energy Web Inf. Technol. **5**(20), 1–5 (2018)
25. Bellairs, J.A., Hasina, R., Agrawal, N.: Tumor DNA: an emerging biomarker in head and neck cancer. Cancer Metastasis Rev. **36**, 515–523 (2017)
26. Awais, M., et al.: Healthcare professional in the loop (HPIL): classification of standard and oral cancer-causing anomalous regions of oral cavity using textural analysis technique in autofluorescence imaging. Sensors (Basel, Switzerland) **20**(20), 5780 (2020)
27. Durgamahanthi, V., Anita Christaline, J., Shirly Edward, A.: GLCM and GLRLM based texture analysis: application to brain cancer diagnosis using histopathology images. In: Dash, S.S., Das, S., Panigrahi, B.K. (eds.) Intelligent Computing and Applications. Advances in Intelligent Systems and Computing, vol. 1172, pp. 691–706. Springer, Singapore (2021)

Machine Learning Based Delta Sigma Modulator Using Memristor for Neuromorphic Computing

Md Noorullah Khan and E. Srinivas[✉]

Anurag University, Hyderabad, India
noorullah.khan@mjcollege.ac.in, srinivasece@anurag.edu.in

Abstract. This paper presents a memristor-based delta sigma modulator that mimics a human brain neuron for artificial neural network training. Memristive synapses are proposed in this work that can be used as weights. In recent works, memristive components have been proposed to be used in ADC and DAC. Such A-to-D converters can be trained using machine learning algorithms to get optimized results for speed and power. This paper describes the design of a 1st-order Sigma-delta ADC using a memristor working with a DC supply voltage of 1.8 V. The proposed ADC can be used for image sensors in augmented reality (AR) and virtual reality (VR) applications. The ADC consumes 24 μW of power at a sampling frequency of 100 kHz with a figure of merit of 3.6 fJ/step. It is ideal for high-resolution, low-frequency applications due to its highly linear characteristics resulting from a single-bit linear quantizer and oversampling techniques.

Keywords: Successive Approximation Register · Artificial Intelligence · Machine Learning · Neuron · Synapse · Delta Sigma modulator

1 Introduction

Recent advancements in image sensing and immersive technologies have led to groundbreaking developments in various domains. This paper focuses on two significant technologies that have garnered substantial attention and are poised to revolutionize their respective fields: Sigma-Delta Modulator Image Sensors and Augmented Reality/Virtual Reality (AR/VR). Image-sensing technology plays a crucial role in numerous applications, such as digital photography, surveillance systems, medical imaging, and industrial inspections. Sigma-Delta Modulator Image Sensors present a promising solution to overcome the limitations of traditional image sensors [1], which include limited dynamic range, high noise levels, and reduced resolution. By employing sigma-delta modulation techniques, these sensors offer enhanced performance characteristics, including higher resolution, improved dynamic range, and superior noise reduction capabilities. Augmented reality (AR) and virtual reality (VR) technologies have gained substantial popularity in recent years, transforming the way we perceive and interact with the digital world. AR enhances the real-world environment by overlaying computer-generated

© The Author(s), under exclusive license to Springer Nature Switzerland AG 2024
S. Satheeskumaran et al. (Eds.): ICICSD 2023, CCIS 2122, pp. 82–95, 2024.
https://doi.org/10.1007/978-3-031-61298-5_7

sensory information [2], while VR immerses users in a completely computer-generated virtual environment. These technologies find applications in various fields, such as gaming, entertainment, education, training, and healthcare, such as medical imaging and radiology [3]. By seamlessly blending virtual and real-world elements, AR/VR technologies offer captivating and interactive experiences that were once considered futuristic concepts.

The basic building blocks of a sigma delta ADC is as shown in Fig. 1, it consists of a Subtractor, followed by an integrator, comparator, D-flip flop to store the digital data and a 1-bit DAC circuit along with digital filter and decimator together comprises Sigma-Delta ADC. The workings of each block are explained in detail in Sect. 2, with respective figures.

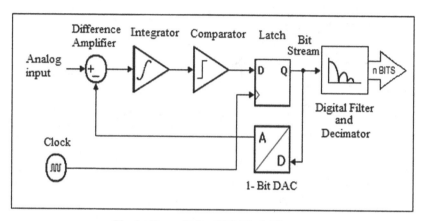

Fig. 1. Sigma-Delta ADC Block Diagram

In the present era, artificial intelligence has become an important part of every application. It can create a great change in the medical field, space exploration, robotics, and computer vision (CV) [3]. CV can be used in machine image processing and machine learning [7]. The analog-to-digital con converter (ADC) is a critical part of every CV system because the inputs to the system coming from the environment via sensors are analogueue, which has to be converted to digital as most AI algorithms are in the digital 8]. In addition to this, speed and power consumption are critical features of the ADC.

ADC's can be implemented with neural architecture using memristors in order to minimize the power consumption since neurons consume very little power in the order of nanowatts. The nervous system consists of neurons with synapses as the key element. Neurons can be excited electrically; they can process and transmit information in the form of electrical signals. Multiple neurons are interconnected to form synapses. A large number of neurons with these interconnections form a neural network, as shown in Fig. 2. A biological neuron consists of three distinct parts: the first is the dendrites, the second is the cell body, and the third is the axon. The junction between the two neurons is called a synapse. Neurons are not physically interconnected to each other; instead, they are separated by a small amount of space called a synaptic cleft.

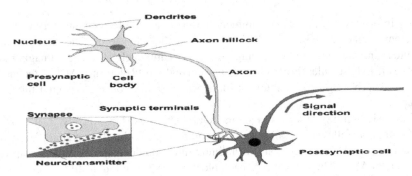

Fig. 2. Biological neuron

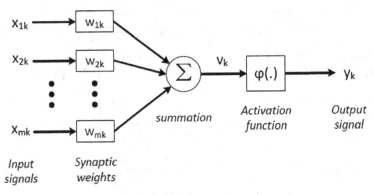

Fig. 3. Artificial neuron model

The biological neuron can be modelled as an artificial neuron, as shown in Fig. 3, where it shows a typical model of the artificial neuron. It consists of three basic elements: first, a set of input synapses that can be multiplied by the synaptic weights; second, a summation unit of the input signals; and third, an activation function to modulate the output signal amplitude. The input signals are first multiplied by the corresponding synaptic weights, as shown in the below Eq. (1)

$$V_K = \sum_{j=1}^{m} W_{jk} X_j \tag{1}$$

where m is the number of neurons, Wjk is the synaptic weight between the j-th neuron and the current neuron k, and Xj is the input signal coming from the j-th neuron.

Vk is the linear summation of the weighted input signals.

The summation of these product terms is converted to an output signal, which is represented by an activation function as shown in Eq. (2).

$$Y_K = \sigma(k) \tag{2}$$

where Yk is the output signal of the kth neuron and Sigma is the activation function, The signum function is the most commonly used for the implementation of ANNs; it is mathematically defined as

$$\sigma(x) = \frac{1}{1 + e^{-\alpha x}} \tag{3}$$

where α is called a slope parameter, for different values of α we get different slopes, as shown in Fig. 4. Multiple neurons are interconnected with each other to form a network called an ANN. The architecture used in Fig. 5 is called a feed-forward architecture, comprising an input layer, a hidden layer, and an output layer. The information is passed from the input to the output layers via hidden layers. The results of the network are given by the neurons of the output layers. Training an artificial neural network is accomplished by the process of adjusting the synaptic weights for a given specific application. Using efficient machine learning algorithms, the weights are adjusted appropriately.

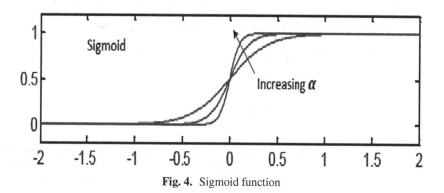

Fig. 4. Sigmoid function

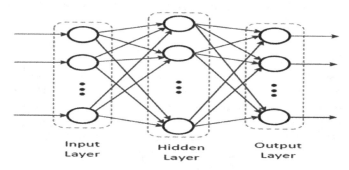

Fig. 5. Feed forward artificial neural network

The organization of the paper is as follows: Sect. 1.1 describes the basic blocks of ADC; Sect. 1.2 describes the neural network concepts; Sect. 1.3 describes the memristor; and Sect. 2 describes the various circuits at schematic levels.

2   Memristor for Artificial Neural Network

The memristor is called the combination of memory and resistor. Memristors are the basic building blocks in the architecture of artificial synapses. In order to complete the gap in circuit theory, the existence of the fourth fundamental passive circuit element was predicted by Prof. Chua in 1971 [8]. The mathematical relationship is shown in Fig. 6.

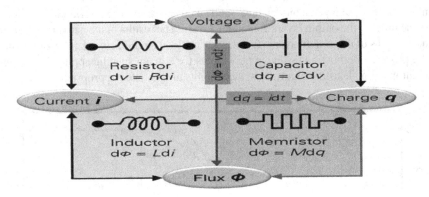

Fig. 6. Memristor

In the emerging electronic devices that use AI exhaustively, the use of memristoris is very essential. Memristors have similar features to biological synapses [13], in storing and dealing with information, and they can also perform computational tasks in the memory itself. Since they consume very little power, in the order of nanowatts, are highly efficient, and have a very small size, they are used in neuromorphic applications,

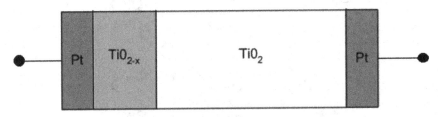

Fig. 7. Memristor structure

Memristors can be made with the help of a material called titanium dioxide, as shown in Fig. 7. They are made by taking two thin layers of titanium dioxide material and placing them with platinum electrodes on either side. The core layer functions like an insulator.

But by doping it with oxygen atoms, it provides two different layers with different resistances; the region that is doped has low resistance and good conductivity when compared with the other. With the removal of the bias voltage, the oxygen vacancies

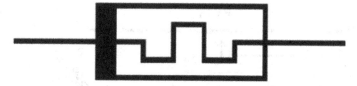

Fig. 8. Memristor symbol

do not shift and the region remains the same, which makes it suitable for use as a non-volatile medium. TiO2 − x represents the doped area, which is oxygen deficient and serves as R_{ON}, and the un-doped area acts as R_{OFF}. The symbol of the memristor is as shown in Fig. 8.

This work uses the VTEAM model for the memristor for its switching characteristics, as shown in the below Eq. (4)

$$\frac{dw(t)}{dt} = \begin{cases} K_{OFF}\left(\frac{V(t)}{V_{OFF}} - 1\right)^{\alpha_{OFF}} f_{OFF}(W), 0 < V_{OFF} < V \\ 0, V_{ON} < V < V_{OFF} \\ K_{ON}\left(\frac{V(t)}{V_{ON}} - 1\right)^{\alpha_{ON}} f_{ON}(W), V < V_{ON} < 0 \end{cases} \tag{4}$$

where V_{on} and V_{off} are threshold voltages, α_{on}, α_{off}, k_{off} and k_{on} are fitting parameters. $F_{off}(w)$ and $f_{on}(w)$ are window functions to restrict w between 0 and 1. The above equation state changes from 0 to 1, where ON state is represented by 0 and off state is represented by 1, Also, the V-I relationship of the VTEAM model is as shown

$$i(t) = \left[R_{ON} + \frac{R_{OFF} - R_{ON}}{W_{OFF} - W_{ON}}(W - W_{ON})\right]^{-1} V(t) \tag{5}$$

R_{ON} and R_{OFF} are the ON and OFF resistances, where w_{on} and w_{off} are the boundaries of the state variable w.

3 Two Stage CMOS Op-Amp Using Memristor

This circuit consists of a differential amplifier in the first stage and a voltage amplifier in stage two, as shown in Fig. 9. The two-stage amplifier offers advantages such as high gain, improved bandwidth, and enhanced performance. Its differential amplifier stage provides amplification and noise rejection, while the voltage amplifier stage further amplifies the signal. With proper biassing and feedback networks, this configuration can be tailored for various analogue circuit applications. Memristors are used in op amps, as shown in Fig. 9, to reduce power dissipation as they consume nanowatts of power.

By using the following gain equation the transistor widths and lengths are designed

$$A = g_m R_D \tag{6}$$

The simulation results of the two-stage op-amp obtained are 75(dB) dB gain, 58.44 phase margin, 48 V/us positive slew rate, and 62 V/us negative slew rate, as shown in Fig. 10.

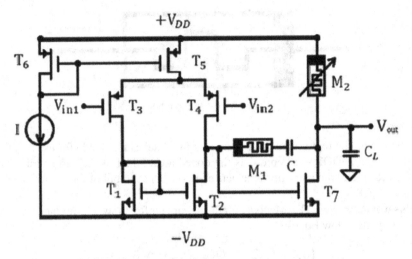

Fig. 9. Schematic of Op-Amp using memristor

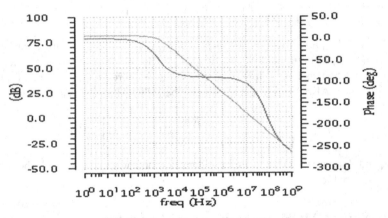

Fig. 10. Simulation results of op-amp

3.1 Subtractor

By using an op-amp in a closed-loop configuration, we can perform the subtraction operation, which subtracts the difference between the two input signals,

$$A = K(V_+ - V_-) \tag{7}$$

The subtractor, as shown in Fig. 11, consists of four resistors arranged in a specific configuration. Two of the resistors are connected in series between the input and the feedback loop, which provides the negative feedback, and the other two resistors serve as a potential divider network where the voltage drop across R2 is used for subtraction. The junction between the two sets of resistors serves as the output node for the error signal. Overall, the subtractor is a fundamental component in the analogue subtractor stage of a $\Sigma\Delta$ modulator. Its purpose is to subtract the feedback signal from the input, generating an error signal that is subsequently processed on the digital platform to achieve high-resolution conversion and noise shaping.

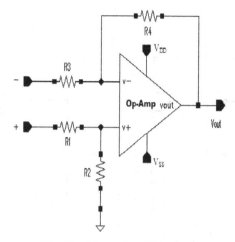

Fig. 11. Schematic of Subtractor

3.2 Integrator

An integrator performs the integration of the signal coming from the output of a subtractor; also, a feedback resistor Rf is used to discharge the capacitor. The practical integrator circuit is shown below in Fig. 12.

$$V_O = -\frac{1}{RC} \int V_i dt \qquad (8)$$

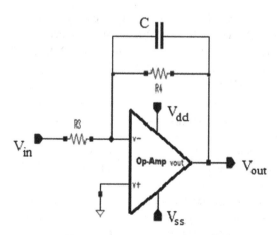

Fig. 12. Schematic of Integrator

3.3 Comparator

The comparator circuit compares the two input signals and specifies the output as greater or lesser, as shown in Fig. 13.

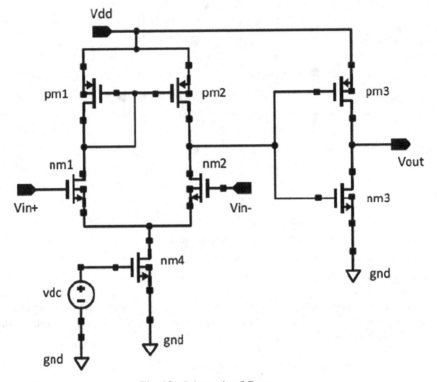

Fig. 13. Schematic of Comparator

In the following design, a 5 mV signal must be resolved using the comparator. The power supply rails are VDD = 1.8 V and VSS = 0 V. We need to maintain a common mode of 0.9 V. That is, the output will swing from 0 V to 1.8 V at a DC of 0.9 when the input signal swings by 5 mV (from 895 mV to 905 mV). The comparator gain must be at least 20,000 (=10 V/5mV) (Table 1).

Table 1. Comparator Design Specifications

Parameter	Value
Supply voltage	1.8 V
Gain	86 dB
Common Mode input	0 to 1 V
Slew Rate	100 V/us
Output Voltage swing	0 to 1.6 v

3.4 D Flip-Flop

In order to provide delayed output and for the sake of sampling purposes, a D flip-flop is added after the comparator, as shown in Fig. 14.

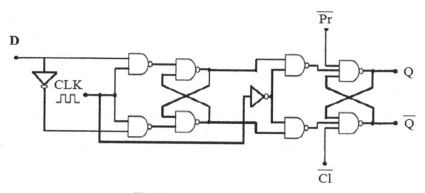

Fig. 14. Schematic of D flip-flop

It changes the comparator output levels from 0 V to 1.8 V. D-flip fop is implemented using master slave configuration. It consists of two separate blocks called master and slave, which are given complementary clocks to each other. When master is on, slave is off, and vice versa. The first block is called master because all four possible inputs can be applied to master, and the second block is called slave because it is given only two possibilities of the input and changes only in response to the master.

3.5 1-Bit DAC

A one-bit DAC is implemented using complementary MOSFET PMOS and NMOS, which converts digital data at the output back to the analogue signal used for calculating the error signal by subtracting it from the reference input signal as shown in Fig. 15.

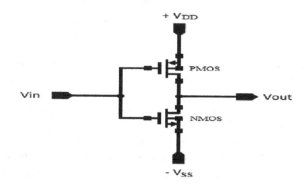

Fig. 15. Schematic of 1-bit DAC

PMOS is on for low voltage, and NMOS is on for high voltage of the input signal for which we get the analogue output.

4 Sigma Delta ADC

The schematic for a sigma-delta modulator is shown in Fig. 16. A sinusoidal signal with a frequency of 400 Hz and a voltage of 500 mV is supplied into the subtractor circuit, and its other terminal is linked to the feedback output signal of a one-bit DAC. The integrator receives the output of this circuit, and its other end is wired to ground. The comparator, whose other two terminals are coupled to clock signals, receives the integrator's output and helps the comparator boost the circuit's speed and sensitivity. The comparator generates the output as previously depicted by comparing the input signal from the integrator with the reference signal. When the input signal passes the reference signal above its value, it will output a positive signal; when it crosses the reference signal, it will output a negative signal.

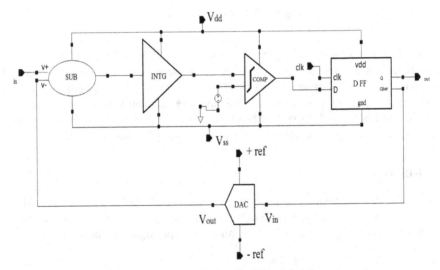

Fig. 16. Schematic of Delta Sigma Modulator

The DAC, which serves as feedback to the subtractor circuit, receives the bit pulse created by the comparator as an input. To produce a series of digital bit streams, this operation is repeated numerous times.

Figure 17 shows the sine wave input to a 1-bit Σ-Δ converter and the respective digital output of the resulting oversampling frequency. Several factors are very important when designing the Σ-Δ converter. The higher the oversampling frequency, the better the output resolution. Therefore, an oversampling frequency of 100 kHz is chosen. The above Fig. 17 shows the simulation output results of a Σ-Δ modulator in a Cadence environment at different stages. For an input of a 400 Hz sinusoidal waveform, with an oversampling frequency of 100 kHz, it can be converted to a digital output. As shown

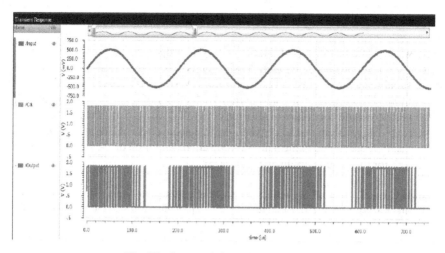

Fig. 17. Output of sigma delta modulator

above, we proposed a Sigma Delta neural network memristor ADC that can be trained for machine learning. Using the stochastic gradient descent (SGD) algorithm, the artificial synapses using memristors [8] can be trained until the ADC achieves optimal parameters of power dissipation and figure-of-merit (FOM). The training algorithm is used to get optimal power consumption and circuit mismatch problems. The performance metrics of various parameters are described in Table 2.

Table 2. ADC performance summary

Characteristic Parameter	[11]	[12]	[13]	This work
Technology	180 nm	90 nm	90 nm	0.18 um
Power supply	1 V	1 V	1 V	1.8 V
Frequency	10 MHz	20 MHz	100 MHz	100 kHz
Resolution	10 bit	9 bits	9 bits	16 bits
FOM	11 (fJ/step)	65 (fJ/step)	18 (fJ/step)	3.6 (fJ/step)
Power	98 uW	290 uW	0.75 mW	24 uW
Reconfigurable	NO	NO	NO	YES

5 Conclusion

In this work, a memristor-based data converter is proposed that solves resistor mismatches in conventional sigma delta ADC circuits. The circuit design is achieved by using a variable memristor instead of fixed values. Verilog-A VTEAM of Cadence software is utilized to obtain the simulation results. A model of the resistor is simulated to get the desired result. Since the memristor occupies a very small area, we get a very high-density design compared to conventional data converters. Also, power dissipation can be greatly reduced due to the nanowatt power consumption of the memristor, which can be used for machine learning applications in AI.

Acknowledgement. The authors wish to express their gratitude to Anurag University and Muffakham Jah College of Engineering and Technology for providing research facilities and technical support.

References

Chiu, Y., Nikolic, B., Gray, P.R.: Scaling of analog-to-digital converters into ultra-deep-dubmicron CMOS. In: Proceedings of the IEEE 2005 Custom Integrated Circuits Conference, pp. 375–382 (2005)

Danial, L., Wainstein, N., Kraus, S., Kvatinsky, S.: Breaking through the speed-power-accuracy tradeoff in ADCs using a memristive neuromorphic architecture. IEEE Trans. Emerg. Topics Comput. Intell. 2(5), 396–409 (2018)

Lewis, S.H.: Optimizing the stage resolution in pipelined, multistage, analog-to-digital converters for video-rate applications. IEEE Trans. Circuits Syst. II: Analog Digit. Signal Process. 39(8), 516–523 (1992)

Lee, C.C., Flynn, M.P.: A SAR-assisted two-stage pipeline ADC. IEEE J. Solid-State Circuits 46(4), 859–869 (2011)

Çikan, N.N., Aksoy, M.: Analog to digital converters performance evaluation using figure of merits in industrial applications. In: European Modelling Symposium, pp. 205–209 (2016)

Neftci, E.O.: Data and power efficient intelligence with neuromorphic learning machines. iScience 5, 52–68 (2018)

Tankimanova, A., James, A.P.: Neural network-based analog-to- digital converters. In: Memristor and Memristive Neural Networks (2018)

Chua, L.: Memristor-the missing circuit element. IEEE Trans. Circuit Theory 18(5), 507–519 (1971)

Danial, L., Kvatinsky, S.: Real-time trainable data converters for general purpose applications. In: Proceedings of the IEEE/ACM International Symposium on Nanoscale Architectures (2018)

Ben Aziza, S., Dzahini, D., Gallin-Martel, L.: A high speed high resolution readout with 14-bits area efficient SAR-ADC adapted for new generations of CMOS image sensors. In: 2015 11th Conference on Ph.D. Research in Microelectronics and Electronics (PRIME), pp. 89–92 (2015)

Liu, C.-C., Chang, S.-J., Huang, G.-Y., Lin, Y.-Z., Huang, C.-M.: A 1 V 11 fJ/conversion-step 10 bit 10 MS/s asynchronous SAR ADC in 0.18 um CMOS. In: IEEE Symposium on VLSI Circuits Diagnosis, pp. 241–242 (2010)

Craninckx, J., Van der Plas, G.: A 65 fJ/conversion-step 0-to-50 MS/s 0-to-0.7 mW 9 b charge-sharing SAR ADC in 90 nm digital CMOS. In: IEEE ISSCC Dig. Tech. Papers, pp. 246–600 (2007)

Lin, Y.-Z., Liu, C.-C., Huang, G.-Y., Shyu, Y.-T., Chang, S.-J.: A 9-bit 150-MS/s 1.53-mW subranged SAR ADC in 90-nm CMOS. In: IEEE Symposium on VLSI Circuits Dignosis, pp. 243–244 (2010)

Correia, A., Barquinha, P., Marques, J., Goe, J.: A high-resolution Δ-modulator ADC with oversampling and noise-shaping for IoT. In: 2018 14th Conference on Ph.D. Research in Microelectronics and Electronics (PRIME), pp. 33–36 (2018)

Garje, K., Kumar, S., Tripathi, A., Maruthi, G., Kumar, M.: A high CMRR, high resolution Bio-ASIC for ECG signals. In: 2016 20th International Symposium on VLSI Design and Test (VDAT), pp. 1–2 (2016)

Danial, L., Wainstein, N., Kraus, S., Kvatinsky, S.: DIDACTIC: a data-intelligent digital-to-analog converter with a trainable integrated circuit using memristors. IEEE J. Emerg. Select. Topics Circuits Syst. **8**(1), 146–158 (2018)

Soudry, D., Di Castro, D., Gal, A., Kolodny, A., Kvatinsky, S.: Memristor-based multilayer neural networks with online gradient descent training. IEEE Trans. Neural Netw. Learn. Syst. **26**(10), 2408–2421 (2015)

Sandrini, J., Attarimashalkoubeh, B., Shahrabi, E., Krawczuk, I., Leblebici, Y.: Effect of metal buffer layer and thermal annealing on HfOx-based ReRAMs. In: 2016 IEEE International Conference on the Science of Electrical Engineering (ICSEE), pp. 1–5 (2016)

An Effective Framework for the Background Removal of Tomato Leaf Disease Using Residual Transformer Network

Alampally Sreedevi[✉] and K. Srinivas[✉]

Computer Science and Engineering, Koneru Lakshmaiah Education Foundation, Aziznagar, Moinabad Road, 500075 Hyderabad, Telangana, India
{sreedevi.a,srirecw9}@klh.edu.in

Abstract. A precise and timely diagnosis of tomato plant diseases is essential for food security and sustainable agriculture. Using photographs of plant foliage and computer vision algorithms, the automated diagnosis of plant diseases has produced encouraging results. However, the accuracy of disease identification algorithms is frequently hindered by the complex and clogged backgrounds of these photographs. To address this issue, a transformer-based network-based classification model for tomato diseases is proposed. Real-time data is used to gather the initial, unprocessed images of tomato plants. Preprocessing was performed in order to eliminate the undesirable image pixels. In addition, the Fully Convolutional Network (FCN) is used to remove photograph backgrounds. The Residual Transformer Network (RTN) is ultimately utilized to classify maladies. Various metrics are employed to validate performance, which is contrasted with more conventional approaches. The RTN model classified tomato leaf diseases successfully. The accuracy analysis for the suggested tomato leaf classification model, RTN, demonstrated 8.04 percent better results than CNN, 6.81 percent better results than Res-net, and 4.44 percent better results than RNN with a higher background removal rate. The results demonstrate that an exceptional categorization rate is achieved to prevent a decline in agricultural output.

Keywords: Tomato Leaves · Disease detection · Background Removal · Fully Convolutional Network · Residual Transformer Network · pattern recognition · computer vision · image processing · computer vision · plant pathology

1 Introduction

The tomato is a widely accessible vegetable variety that is offered on a worldwide scale. However, it has been shown that the productivity of tomato plants is adversely influenced by the presence of numerous leaf diseases [1]. The drop in tomato crop output and subsequent economic loss to farmers may be attributed to the infection of tomato leaves. Therefore, the identification of leaf diseases in tomato plants is closely associated with the agricultural sector's economic endeavors [2]. Most conventional plant disease detection techniques need several analyses conducted by persons to assess the disease-damaged

© The Author(s), under exclusive license to Springer Nature Switzerland AG 2024
S. Satheeskumaran et al. (Eds.): ICICSD 2023, CCIS 2122, pp. 96–109, 2024.
https://doi.org/10.1007/978-3-031-61298-5_8

leaf using chemical or visual observation in the affected area. This approach is associated with a low identification rate and poor reliability due to the potential for human error [3]. Similarly, the farmers possess little expertise, and the absence of skilled specialists to accurately identify the illness negatively impacts the pace of crop production. Therefore, the lack of attention to detail results in several challenges in relation to global food security and has additional negative consequences for stakeholders involved in tomato production [4]. Early disease detection and classification models play a crucial role in assisting farmers in addressing agricultural concerns [5].

Real-time application has a significant role in improving disease detection and classification rates in the field of agriculture [6]. These applications need the implementation of low-latency solutions in their devices, as well as the reduction of computing power and storage space [7]. The primary focus of specialists in the field is to develop lightweight models using a restricted number of samples afflicted by the illness. Furthermore, researchers have made advancements in the development of plant disease detection models by using conventional machine learning techniques [8]. Deep learning models have shown significant efficacy in the field of agriculture, providing users with precise and reliable results. In contemporary times, the Convolutional Neural Network (CNN) has emerged as a crucial instrument for the classification of various tasks. It has shown its efficacy in extracting pertinent information from images autonomously, without the need for user intervention [9]. The conventional deep learning approach for tomato plant leaf disease encounters several challenges that significantly impact its effectiveness. In the first stages, the occurrence of picture noises may be seen during the acquisition process. Furthermore, the subsequent steps of processing and transmission provide further complication in acquiring the disease-related information [10]. In order to address the many constraints inherent in the traditional method of classifying tomato plant leaf diseases, a new framework for classifying these diseases has been devised.

This framework utilizes a network methodology and incorporates a background removal mechanism.

- The current framework demonstrates several contributions pertaining to the proposed method for the management of leaf diseases in tomato plants.
- The objective of this study is to develop an innovative methodology for classifying tomato leaf diseases using deep learning models. The aim is to accurately identify and classify diseases in tomato plants at an early stage, hence improving the overall yield production rate.
- In order to improve the accuracy of leaf disease classification, the FCN (Fully Convolutional Network) method is used to remove the background from the picture, resulting in an upgraded image with the background eliminated.
- In order to categorize tomato leaf diseases, it is recommended to use the RTN approach, since it has been shown to provide a higher rate of accuracy in classifying tomato leaf diseases compared to current models.
- Multiple studies are used to assess the classification rate of the proposed tomato leaf disease classification model in comparison to traditional classifiers.

The subsequent sections pertaining to the suggested methodology for classifying tomato leaf diseases will be further discussed in the following paragraphs. The studies associated with the traditional classification approach for tomato leaf diseases are described in Sect. 2. The discussion of the image background reduction model using Fully Convolutional Networks (FCN) is presented in Sect. 3. Section 4 presents a strategy for classifying tomato leaf diseases that has shown effectiveness. The presentation of the findings and discussion are in Sect. 5, while the conclusion is given in Sect. 6.

2 Literature Survey

2.1 Related Works

Zhou et al. (2021) proposed a modified Residual Dense Network (RDN) for the purpose of tomato leaf disease identification. The deep learning model, which incorporates the advancements introduced in the Deep Residual Network (DRN), aims to reduce the number of parameters required during training in order to improve the accuracy rate. The researchers used the existing Residual Dense Network (RDN) to improve the resolution of images by modifying the network architecture. Ultimately, the suggested Registered Dietitian Nutritionist (RDN) achieved a higher percentage of accurate categorization for plant leaf diseases compared to the currently used methods.

Ahmed et al. (2022) have devised a low-complexity learning model for the purpose of disease detection in tomato leaves. The proposed methodology used an effective pre-processing strategy to enhance the lighting correction in the leaf picture. The characteristics of the initial procedures were obtained by the integration of fused methods, which include a pre-trained mobile net structure with a classifier to enhance the efficiency of prediction. In this study, the occurrence of data leakages was mitigated by the use of runtime augmentation techniques. The results of the experimental study demonstrated that the proposed leaf disease classification approach achieved a much better accuracy rate compared to traditional techniques.

Anandhakrishnan and Jaisakthi (2022) have proposed the use of a Deep Convolutional Neural Network (DCNN) in order to enhance the accuracy and reduce the reaction time in the detection of tomato leaf diseases [13]. The analysis of the created model included the use of substantial quantities of data. In order to get a favorable result, both training data and testing data were used. Therefore, the DCNN approach advocated for the classification model of tomato plant leaf diseases shown a significantly superior performance rate compared to traditional models.

Zhang et al. (2023) have proposed an asymptotic method to reduce the interference caused by noise in images, aiming to simplify the process of obtaining deep features for improved detection. Additionally, the Multi-channel Automatic Orientation Recurrent Attention Network (M-AORANet) was devised with the objective of capturing a greater number of disease-related characteristics. Therefore, the suggested methodologies successfully addressed the problem of similarity across different classes and resulted in a higher accuracy in classifying leaf diseases compared to the traditional approach.

In the year 2022, Kaushik et al. [15] presented a novel approach for detecting diseases in tomato plants, which they dubbed TomFusioNet. During the feature extraction phase, the use of the late-fusion technique was employed to achieve a high success rate. The model that was constructed consisted of a fused mixture of DeepPred and DeepRec algorithms, which was used for the purpose of disease detection in crops. The hyperparameters of DeepPred were improved using a genetic technique. Additionally, a background noise reduction technique based on Hue, Saturation, and Value (HSV) was developed in order to improve the efficiency of performance in Hue. Therefore, the proposed model demonstrated improved classification performance compared to traditional strategies in terms of accuracy.

2.2 Problem Statement

The issue of food security in tomato plants has garnered increased global attention in recent times. However, several illnesses acquired by tomato plants have been shown to significantly decrease the pace of production. Numerous models for classifying tomato leaf diseases have been created; nevertheless, it is imperative that such systems exhibit enhanced accuracy, swifter processing capabilities, and computational affordability in order to effectively assist farmers. Table 1 displays the many difficulties and traits that are included in the conventional methods for classifying tomato leaf diseases. The RDN [11] algorithm demonstrates efficient classification capabilities while using a limited number of parameters. Additionally, it efficiently addresses the problem of gradient disappearance. However, it is important to address the temporal complexity concerns in order to enhance the efficiency of the process. The MobileNet V2 architecture [12] demonstrates superior accuracy in leaf disease classification and effectively addresses challenges related to data leakage and class imbalance. However, the system exhibited limitations in its ability to accurately diagnose a wider range of illnesses based on a single leaf. The Deep Convolutional Neural Network (DCNN) has been shown to effectively reduce reaction time and enhance the accuracy of leaf disease classification [13]. Furthermore, the system had significant challenges when presented with a picture with a higher level of noise throughout the execution process. The $M - AORANet$ [14] algorithm aims to mitigate the impact of noise interference in images and simplify the process of extracting leaf disease information. Nevertheless, there is a need to improve the accuracy of the categorization rate. TomFusioNet [15] is capable of automatically detecting the kind of illness and classifying its classes. Furthermore, it has a higher level of accuracy in detecting leaf diseases during the early stages. Background removal is a necessary preprocessing step for accurately assessing diseases by cleaning and preparing the input data. Machine learning algorithms have the capability to accurately detect and categorize illness patterns within a narrower and more regulated framework.

Table 1. Features and challenges of conventional Tomato leaf disease classification models

Author [citation]	Methodology	Features	Challenges
Zhou et al. [11]	RDN	• The classification task is executed efficiently and effectively, using a limited number of parameters • The issue of the gradient fading is essentially resolved	• In order to enhance the efficiency of the process, it is essential to address the time complexity concerns
Ahmed et al. [12]	MobileNet V2	• The accuracy of leaf disease classification was improved • The proposed method effectively addresses the issues of data leakage and class imbalance, resulting in minimized occurrences of these problems	• The system exhibited limitations in its ability to accurately diagnose a broader range of illnesses based on a single leaf
Anandhakrishnan and Jaisakthi [13]	DCNN	• The proposed approach effectively reduces response time and enhances the accuracy of leaf disease categorization	• The performance of the system was significantly impacted when presented with a picture with a higher level of noise
Zhang et al. [14]	M − AORANet	• The proposed method aims to mitigate the impact of noise interference in the picture and simplify the process of extracting leaf disease information	• The objective is to improve the accuracy of the categorization rate
Kaushik et al. [15]	TomFusioNet	• It automatically detects the disease type and classifies their classes • It detects the leaf disease more accurately in the initial phases	• It utilized only limited data for the analysis and suffered in some cases

3 Tomato Leaf Disease Classification with Deep Learning Using Background Removal

3.1 Dataset Information

The tomato leaf images are initially utilized for analysis and obtained from real-world situations in order to verify the effectiveness of the suggested model for classifying tomato plant leaves. The photos in the collection were obtained from the Kuntloor village, located in Hayathnagar, Hyderabad. The collection contains around 2000 example photos including many classifications, including Tuta absoluta, fungal-leaf Early blight, healthy leaf, leaf serpentine miner, fungal-leaf gray mold, and tobacco caterpillar leaf damage. The acquired input real-world tomato leaf images are offered as IMG_p^{sp} where $p = 1, 2, 3 ..., P$. The term P denotes total dataset.

3.2 Developed Model

A novel approach is proposed for identifying tomato plant leaf diseases, using heuristic models and deep learning approaches to accurately identify the specific disease variant affecting the crop during its first stages. During the first stage, photos depicting the tomato leaf disease are obtained from real-time scenarios and thereafter sent to the image pre-processing area. In this approach, the pictures undergo pre-processing using the median blur method. The resulting pre-processed image is then used as input for the image background removal step. During this step, the removal of picture backdrops is achieved by using the Fully Convolutional Network (FCN) approach. Moreover, the background-removed pictures obtained are afterwards used in the leaf disease classification domain. Tomato plant leaf diseases are categorized according to RTN systems. Therefore, the tomato leaf disease classification framework that is proposed achieved a higher rate of leaf disease classification compared to the traditional schemes. The model constructed for this purpose is shown in Fig. 1.

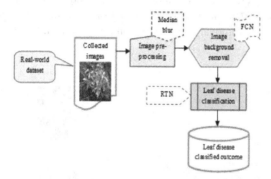

Fig. 1. A schematic depiction of the Tomato Plant Leaf Disease Classification Framework including background removal methods.

3.3 Median Blur-Based Image Pre-processing

Initially, the plant leaf pictures $IMG_p{}^{sp}$ collected from the dataset are sent as input to the pre-processing stage. The use of the median blur method is employed for the purpose of pre-processing the picture depicting plant leaf disease. The median blur approach is a commonly used non-linear digital filtering method that is mostly applied for the purpose of noise reduction in images. The median blur technique operates by selecting the median value from the set of intensities inside a given timeframe. The algorithm works by operating on a pixel-by-pixel basis and replacing the value of each pixel with the median value of its neighboring pixels. The neighbor pattern, also known as the window, is applied to the whole picture in a pixel-by-pixel manner. The practice of minimizing this kind of noise is widely recognized as a fundamental pre-processing step in order to improve the overall result of the pre-processing stage. The pre-processed images obtained using median blur, denoted as $IMGq^{MD}$, are then used for the background removal stage.

4 The Concept of Background Removal in Tomato Leaf Disease Classification Using Deep Learning

4.1 Background Removal Using FCN

The pre-processed photos $IMGq^{MD}$ are used as input for the background removal step. In this study, the background removal procedure is executed on the picture using the Fully Convolutional Network (FCN) approach.

The down-sampling layer of the Fully Convolutional Network (FCN) used a total of 19 layers for the purpose of analysis. Down-sampling approaches are used for analysis due to their improved effectiveness and ability to shorten the training duration. The phase up-sampling is regarded as a crucial component of the Fully Convolutional Network (FCN) due to its effective reversal of the down-sampling process used in dense prediction. In the area of up-sampling, the integration of both local and global information occurs via the incorporation of specialized layers derived from deconvolutional and convolutional layers. This process yields the elimination of background noise as the resulting output. Moreover, the picture that has undergone backdrop removal is referred to as Fig. 2. Depicts the background removal model based on Fully Convolutional Networks (FCN).

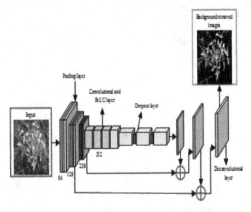

Fig. 2. Diagrammatic representation of the FCN-based background removal model

4.2 Residual Transformer Network-Based Disease Classification

The leaf image classification phase utilizes the FCN-based backdrop picture as its input. The RTN approach is used in the leaf disease classification phase to achieve efficient categorization of leaf diseases. The development of the RTN is rooted on the integration of Resnet and transformer methodologies. The Resnet network successfully educated the layer without increasing the error rate. It effectively mitigates the problem of gradient vanishing. However, the current analysis lacks sufficient strength to establish a correlation due to the absence of data pertaining to the missing variable. Moreover, the use of a transformer is justified due to its great efficiency in handling a wide frequency range. Additionally, the implementation of a transformer is believed to be relatively easier, and they are interconnected in a reverse configuration. There is a need for more spatial capacity, and the object in question has a substantial and unwieldy physical dimension. Therefore, in order to address the many intricacies encountered in Resnet and the transformer, a novel amalgamation called RTN has been devised. The RTN model that has been designed consists of two crucial layers: the transformer layer and the shortcut connection. The transformer assumes a significant role since it has the capability to establish temporal dependencies within the incoming data. The transformer model utilizes linear layers to handle input data in several traffic modes and successfully establishes correlations across different traffic nodes. Therefore, the implemented Recurrent Transition Network (RTN) successfully achieved accurate classification of leaf diseases. Figure 3. Illustrates the visual depiction of the leaf disease classification approach based on the Recursive Transition Network (RTN).

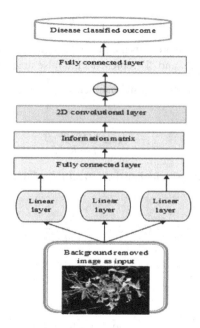

Fig. 3. The pictorial view of the RTN-based leaf disease classification method

5   Result and Discussion

5.1   Experimental Setup

The present study included the use of Python software to develop a unique framework for the classification of tomato leaf diseases. Multiple investigations were undertaken to verify the efficacy of the suggested methodology for categorizing illnesses affecting tomato leaves. The assessment of the effectiveness of the suggested approach for the categorization of tomato leaf diseases was conducted via a comparative analysis with previous classification methods used for tomato leaf diseases. The research conducted a comparative analysis between a novel approach for identifying tomato leaf diseases and many established classification models, such as CNN [17], Resnet [18], and RNN [19], with the aim of obtaining accurate classification outcomes.

5.2 Performance Metrics

The proposed framework for classifying tomato plant leaves is based on many qualitative metrics, which are described below.

(a) The calculation of metrics associated with a certain gain is referred to as accuracy and it stated in Eq. (1).

$$AY = \frac{(k+j)}{(k+j+h+g)} \tag{1}$$

In this phase, the term "j" represents the true negative, "h" signifies the false positive, "k" presents the true positive, and "l" represents the false negative values.

(b) Specificity refers to the accurate establishment of negative propositions as shown in Eq. (2).

$$Ik = \frac{j}{j+h} \tag{2}$$

(c) F1-score is a metric used to calculate the correctness of the whole analysis, as shown in Eq. (3).

$$Rf = 2 \times \frac{2k}{2k+h+g} \tag{3}$$

(d) The calculation of binary quality categories of the observation, referred to as MCC, is represented by Eq. (4).

$$Qa = \frac{k \times j - h \times g}{\sqrt{(k+h)(k+g)(j+k)(j+g)}} \tag{4}$$

(e) The variable FNR represents the proportion of positive cases that resulted in negative observations in the various analyses presented in Eq. (5).

$$Rf = \frac{g}{g+k} \tag{5}$$

(f) The net present value (NPV) is calculated as the aggregate value of all plants that are free from disease during the study, as expressed in Eq. (6).

$$Rd = \frac{j}{j+g} \tag{6}$$

(g) The false positive rate (FPR) is defined as the proportion of negative events that are incorrectly classified as positive, as represented by the Eq. (7).

$$Ws = \frac{h}{h+j} \tag{7}$$

(h) The sensitivity refers to the proportion of true positive cases that are correctly identified, as described in Eq. (8).

$$pl = \frac{k}{k+g} \tag{8}$$

(i) Precision is defined as the ratio of the number of true positive predictions to the total number of positive predictions, as represented by Eq. (9).

$$PRS = \frac{k}{k+h} \tag{9}$$

Image Description-Disease name	Background removed image
Healthy leaf	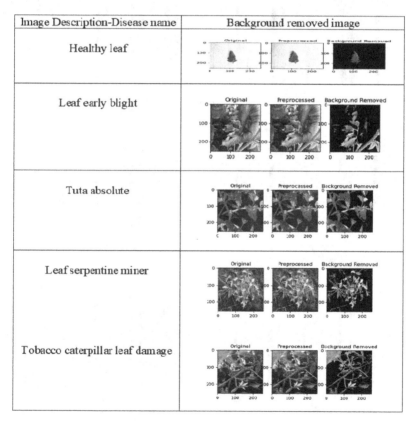
Leaf early blight	
Tuta absolute	
Leaf serpentine miner	
Tobacco caterpillar leaf damage	

Fig. 4. The resulting picture output after background removal in the framework for classifying tomato leaf diseases.

5.3 Resultant Background Removed Images

The tomato leaf disease classification model in Fig. 4 demonstrates the use of FCN-based background removal pictures. This study presents a collection of original photos that have undergone preprocessing and background removal. The images belong to five distinct groups.

5.4 Analysis of Initiated Model with Conventional Classifiers

Figure 5 illustrates the various studies conducted inside the advanced framework for classifying tomato leaf diseases, as compared to conventional methodologies. The suggested tomato leaf classification model, RTN, exhibited an improvement of 8.04% compared

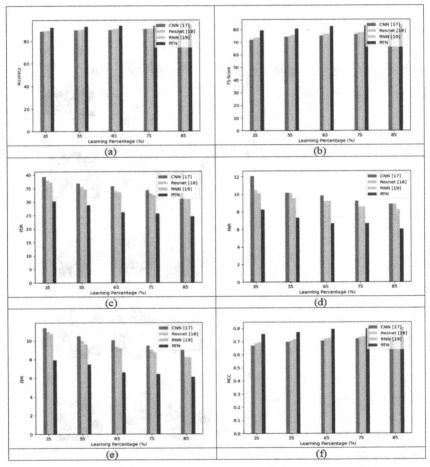

Fig. 5. Analysis of the developed tomato leaf disease classification model with background removal over existing classifiers with (a) Accuracy, (b) F1-score, (c) FDR, (d) FNR, (e) FPR, (f) MCC, (g) NVP and (h) Precision

to CNN, 6.81% compared to Resnet, and 4.445% compared to RNN, as seen in the accuracy study. Therefore, the proposed framework for classifying tomato leaf diseases, which includes background removal, achieved a higher level of accuracy in classifying leaf diseases compared to traditional methodologies.

5.5 Performance Analysis on the Recommended Model

The performance analysis of the tomato leaf disease classification model, as proposed, is shown in Table 2. The tomato leaf disease classification model underwent many studies, and the proposed RTN approach demonstrated improved performance compared to the current technique. The conducted sensitivity study in the first tomato leaf classification resulted in an improved leaf disease classification rate of 3.13%, 3.14%, and 2.42% compared to conventional models such as CNN, Resnet, and RNN, respectively. Therefore, the suggested Recurrent Transition Network (RTN) with an enhanced background removal rate achieved a higher rate of successful classification of leaf diseases compared to traditional methods.

Table 2. Performance analysis of the An Effective Framework for the Background Removal of Tomato Leaf Disease using Residual Transformer Network

Performance measures	CNN [17]	RESNET [18]	RNN [19]	RTN (Proposed)
Accuracy	91.85	91.96	92.31	93.92
Sensitivity	91.54	91.67	92.76	93.97
Specificity	90.93	91.87	91.86	93.92
Precision	66.72	68.97	69.17	75.61
FPR	9.09	8.20	8.20	6.11
FNR	8.90	8.90	8.34	6.09
NPV	98.17	98.19	98.27	98.76
FDR	33.29	31.09	30.89	24.53
F1-Score	77.16	78.57	78.91	83.81
MCC	72.97	74.63	75.14	80.81

6   Conclusion

The implementation of a novel framework for classifying tomato leaf diseases, using deep learning techniques, resulted in an improved rate of classification for leaf diseases in plants during their first stages. Initially, a collection of sick leaf photos was provided as the input to the image pre-processing area. The pre-processed pictures were obtained using the median blur method and afterwards utilized in the image background removal step. The FCN approach was used to eliminate the backdrop of the picture during this step. Subsequently, the background-removed photos obtained were inputted into the area responsible for classifying tomato leaf diseases. The RTN model successfully conducted the categorization of tomato leaf diseases, demonstrating its effectiveness. The accuracy analysis conducted in the proposed tomato leaf classification model, RTN, demonstrated an improvement of 8.04% compared to CNN, 6.81% compared to Resnet, and 4.445% compared to RNN, with an increased background removal rate. Therefore, the proposed framework for classifying tomato leaf diseases, which includes background reduction, achieves a higher rate of disease classification compared to current methodologies.

References

1. Yang, G., Chen, G., He, Y., Yan, Z., Guo, Y., Ding, J.: Self-supervised collaborative multi-network for fine-grained visual categorization of tomato diseases. IEEE Access **8**, 211912–211923 (2020)
2. Schor, N., Bechar, A., Ignat, T., Dombrovsky, A., Elad, Y., Berman, S.: Robotic disease detection in greenhouses: combined detection of powdery mildew and tomato spotted wilt virus. IEEE Robot. Autom. Lett. **1**(1), 354–360 (2016)
3. Zhang, Y., Song, C., Zhang, D.: Deep learning-based object detection improvement for tomato disease. IEEE Access **8**, 56607–56614 (2020)
4. Wu, Q., Chen, Y., Meng, J.: DCGAN-based data augmentation for tomato leaf disease identification. IEEE Access **8**, 98716–98728 (2020)
5. Liu, J., Wang, X.: Early recognition of tomato gray leaf spot disease based on MobileNetv2-YOLOv3 model. Plant Methods **16**, 83 (2020). https://doi.org/10.1186/s13007-020-00624-2
6. Nawaz, M., et al.: A robust deep learning approach for tomato plant leaf disease localization and classification. Sci. Rep. **12**, 18568 (2022)
7. Moussafir, M., Chaibi, H., Saadane, R., Chehri, A., El Rharras, A., Jeon, G.: Design of efficient techniques for tomato leaf disease detection using genetic algorithm-based and deep neural networks. Plant Soil **479**, 251–266 (2022)
8. Ahmed, N., Zaidi, S.S.E.A., Amin, I., Scheffler, B.E., Mansoor, S.:"Tomato leaf curl Oman virus and associated Betasatellite causing leaf curl disease in tomato in Pakistan. Eur. J. Plant Pathol. **160**, 249–257 (2021)
9. Sayed, S.S., et al.: Association of tomato leaf curl Sudan virus with leaf curl disease of tomato in Jeddah, Saudi Arabia. VirusDisease 27, 145–153 (2016)
10. Ashwathappa, K.V., et al.: Association of Tomato leaf curl Karnataka virus and satellites with enation leaf curl disease of Pseuderanthemum reticulatum (Radlk.) a new ornamental host for begomovirus infecting tomato in India. Indian Phytopathol. **74**, 1065–1073 (2021)
11. Zhou, C., Zhou, S., Xing, J., Song, J.: Tomato leaf disease identification by restructured deep residual dense network. IEEE Access **9**, 28822–28831 (2021)
12. Ahmed, S., Hasan, M.B., Ahmed, T., Sony, M.R.K., Kabir, M.H.: Less is more: lighter and faster deep neural architecture for tomato leaf disease classification. IEEE Access **10**, 68868–68884 (2022)
13. Anandhakrishnan, T., Jaisakthi, S.M.: Deep convolutional neural networks for image-based tomato leaf disease detection. Sustain. Chem. Pharmacy **30**, 100793 (2022)
14. Zhang, Y., Huanga, S., Zhou, G., Yahui, H., Lic, L.: Identification of tomato leaf diseases based on multi-channel automatic orientation recurrent attention network. Comput. Electron. Agric. **205**, 107605 (2023)
15. Kaushik, H., Khanna, A., Singh, D., Kaur, M., Lee, H.N.:"TomFusioNet: a tomato crop analysis framework for mobile applications using the multi-objective optimization based late fusion of deep models and background elimination. Appl. Soft Comput. **133**, 109898 (2023)
16. Yang, X., Li, H., Yu, Y., Luo, X., Huang, T., Yang, X.: Automatic pixel-level crack detection and measurement using fully convolutional network, 29 August 2018

17. Daanouni, O., Cherradi, B., Tmiri, A.: NSL-MHA-CNN: a novel CNN architecture for robust diabetic retinopathy prediction against adversarial attacks. IEEE Access **10**, 103987–103999 (2022)
18. Song, S., Lam, J.C.K., Han, Y., Li, V.O.K.: ResNet-LSTM for real-time PM2.5 and PM_{10} estimation using sequential smartphone images. IEEE Access **8**, 220069–220082 (2020)
19. Mansouri, M., Dhibi, K., Hajji, M., Bouzara, K., Nounou, H., Nounou, M.: Interval-valued reduced RNN for fault detection and diagnosis for wind energy conversion systems. IEEE Sens. J. **22**(13), 13581–13588 (2022)

An Intelligent Ensemble Architecture to Accurately Predict Housing Price for Smart Cities

K. Sudheer Reddy(✉) ⓘ, Niteesha Sharma ⓘ, T. Ashalatha ⓘ, and B. Ravi Raju ⓘ

Department of Information Technology, Anurag University, Hyderabad, India
sudheercse@gmail.com

Abstract. Accurate housing price prediction is vital for various real estate applications. This paper presents a comprehensive study on housing price prediction of smart cities, evaluating the performance of individual models and introducing an ensemble approach. The study investigates Linear Regression, Random Forest, Gradient Boost, and XGBoost, while introducing a novel ensemble model combining Gradient Boosting and a Feedforward Neural Network. Using a diverse dataset of housing attributes, we preprocess and engineer features for improved predictive performance. To access the accuracy and explanatory capacity of the model, metrics such as mean squared error (MSE), Mean absolute error (MAE) and R-Squared are utilized. Results indicate that the ensemble model achieves predictive accuracy compared to XGBoost, demonstrating competitive MSE, high R-squared, and low MAE values. Our findings underscore the value of ensemble methods in housing price prediction. The ensemble model's success, alongside the performance of individual models, contributes to informed decision-making for real estate professionals and policymakers. This study advances the discourse on predictive modelling within housing economics, emphasizing the efficacy of ensemble techniques in capturing complex price trends. This research paves the way for further exploration of advanced ensemble methods and feature engineering strategies, offering a foundation for accurate housing price prediction and its implications for real-world applications.

The problem of accurately predicting housing prices in smart cities has persisted, stemming from the limitations of existing models to capture complex price trends and provide reliable decision-making insights. This study aims to address these limitations by rigorously evaluating the performance of individual models, proposing an innovative ensemble approach (GBR + NN), and ultimately contributing to the advancement of predictive modelling within the dynamic landscape of housing economics and urban planning. The suggested model outperforms the conventional model in terms of performance indicators. In this work, the Mean Absolute Error is 105.44, R-squared is 0.99 and Mean Squared Error is 26.63.

Keywords: Housing Price Prediction · Ensemble Modelling · Smart cities · Linear Regression · Random Forest · Gradient Boosting · Feedforward Neural Network

© The Author(s), under exclusive license to Springer Nature Switzerland AG 2024
S. Satheeskumaran et al. (Eds.): ICICSD 2023, CCIS 2122, pp. 110–122, 2024.
https://doi.org/10.1007/978-3-031-61298-5_9

1   Introduction

Accurate housing price prediction holds significant implications for real estate professionals, urban planners, and policymakers alike. The ability to forecast housing prices of smart cities with precision facilitates informed decision-making, Opinion making [1] in various contexts, from property investment to urban development strategies. As the housing market continues to evolve, the demand for robust predictive models becomes increasingly pronounced [2]. In response to this demand, this paper embarks on an in-depth exploration of housing price prediction, employing a comprehensive approach that embraces both individual models and ensemble techniques [3]. The study delves into the performance of widely used models, including Linear Regression [4], Random Forest [5], Gradient Boost [6, 7], and XGBoost [8], while introducing an innovative ensemble model that combines the strengths of Gradient Boosting and a Feedforward Neural Network.

The main goal of this study is to not only assess predictive accuracy of these models but also to introduce a novel ensemble approach of housing price prediction in smart cities that capitalizes on their complementary attributes [9]. By integrating diverse models, the ensemble technique aims to extract intricate patterns and relationships within housing data, ultimately enhancing predictive power. Through a meticulous preprocessing and feature engineering process [10], we ensure that the dataset underpinning this study is robust and reflective of real-world housing attributes. The evaluation of model performance employs key metrics, including MSE [9], R-squared, and MAE, to measure accuracy, explanatory strength, and prediction errors [9].

The findings unveil the potential of ensemble methods in advancing the field of housing price prediction. The ensemble model is comparable with accuracy to cutting edge techniques, alongside the performance of individual models, underscores its utility in forecasting complex market trends. The subsequent sections delve into the models' evaluation, comparative analysis, and implications for decision-makers across the real estate landscape.

2   The Literature Review

Chen and Guestrin's introduces XGBoost, an optimized version of Gradient Boosting. The authors showcase the algorithm's scalability, regularization techniques, and parallel processing capabilities, leading to improved predictive accuracy across diverse datasets [11]. Kong et al. propose an innovative hybrid model that combines Gradient Boosting and a Feedforward Neural Network for housing price prediction. The paper demonstrates the effectiveness of synergizing these two techniques, achieving competitive predictive accuracy and capturing intricate housing market dynamics [12].

A thorough assessment by Winky K.O. Ho et al. examines the state of machine learning-based house price prediction. The writers examine several approaches, feature selection plans, and preprocessing methods to provide insight into the state of the field and the significance of predictive model selection.[13]. H. Prakash et al. present a comparison of techniques for predicting house prices, investigating the performance of different algorithms. The paper offers insights into the predictive capabilities of diverse models and their applicability to real-world housing datasets [14].

Xiao et al.'s comprehensive survey explores predictive analytics' applications in the real estate domain. The authors highlight the use of big data in real estate to solve the challenge of accurately analyzing housing price prediction for decision-making and examine various methods and techniques used in the field [15]. Grybauskas A et al. conduct a comprehensive analysis of predictive analytics' role in understanding housing market dynamics. The paper delves into the implications of predictive modelling for housing market trends, offering insights for real estate professionals and policymakers [16].

P. Y. Wang et.al recent study introduces a deep learning model that incorporates feature representation learning for house price prediction. By automatically learning relevant features, the model enhances predictive accuracy, contributing to the evolving field of deep learning in real estate prediction [16]. Thamarai, M et al. propose a hybrid approach that combines multiple factors with a decision tree model for housing price prediction. Their study showcases the value of integrating diverse information sources to improve predictive performance [17]. The comparison of performance metrics on various models are shown in the Table 1. In the table various models are housing prices are predicted based on the various datasets selected by the authors [19, 20].

Table 1. Comparison of performance metrics on various models used for the prediction of Housing price.

S. No	Performance Metrics	Survey on Predictive Models for Housing Prices			
		Linear Regression	Random Forest	Gradient Boost	XGBoost
1	MAE [19, 20]	0.023	0.022	0.0222	0.018
2	MSE [19, 20]	0.001	0.001	0.001	0.001
3	R Squared [19, 20]	0.040	0.042	0.0368	0.0405

3 Methodology

3.1 Research Data Flow Diagram

Figure 1 illustrates the fundamental methodology used in the evaluation process, wherein the model selection is critical to the accuracy values.

3.2 Data Source and Selection

For the experiment we used the Paris Housing Classification dataset [18]. The dataset is created from the imaginary data of house prices in an urban environment-Paris. There are about 18 attributes and all the attributes are numeric variables. The dataset provides the in-depth information regarding the various like square meters, no. of rooms, floors, made year, garage, basement and many more as shown in Table 2. We extracted 15,135

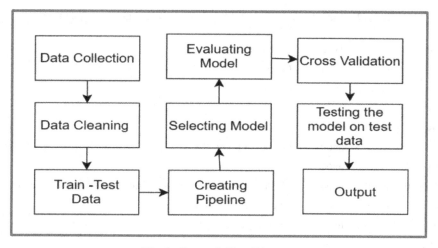

Fig. 1. Research Flow Diagram

records including 18 attributes. Before developing a regression model exploratory data analysis is performed. As the model uses a supervised learning method there is a need of splitting the dataset into training and testing dataset.80% of the dataset is for training and remaining 20% is for testing (Fig. 2).

Visualization of Dataset:

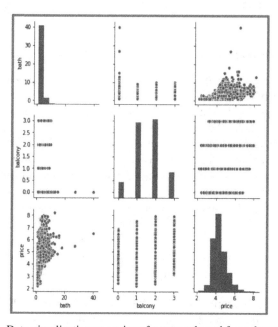

Fig.2. Data visualization on various features selected from dataset [19]

Table 2. Features description of the Paris housing classification dataset.

Name	Type	Description
squareMeters	int64	Units
numberOfRooms	int64	Count
hasYard	int64	Yes/No
hasPool	int64	Yes/No
floors	int64	number of floors
cityCode	int64	zip code
cityPartRange	int64	The higher the range, the more exclusive the neighborhood is
numPrevOwners	int64	number of previous owners
made	int64	Made in year
isNewBuilt	int64	Yes/No
hasStormProtector	int64	Yes/No
basement	int64	basement square meters
attic	int64	attic square meteres
garage	int64	garage size
hasGuestRoom	int64	number of guest rooms
price	float64	predicted value
category	category	Luxury/Normal

3.3 Data Preprocessing

In data preprocessing phase the dataset was cleaned using the feature engineering processes mentioned below:

Firstly, we categorize the dataset depending upon the type of data and then we calculate the number of them. We check if there are any categorical variables present and if there are any, we transform them to numerical using one Hot Encoding. In the above dataset we have the column ["category-Basic, Luxury"] as a categorical variable hence we convert it into numerical value. Remove the attributes cityPartRange, hasStorage-Room, attic due to their ambiguity. The dataset was checked for outliers for the attribute price by checking lower bound (LB) and upper bound (UB) as shown in Eqs. 1 and 2:

$$LB = Q1 - 1.5 * IQR \tag{1}$$

$$UB = Q3 + 1.5 * IQR \tag{2}$$

where Q1 is the 25th Percentile and Q3 is the 75th Percentile and IQR = Q3-Q1.

After applying Eqs. 1 and 2 to the price column there were no potential outliers found the result of it is as shown in the Fig. 3.

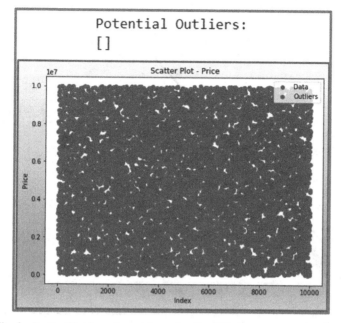

Fig. 3. Scatter Plot for price attribute showing NO OUTLIERS are present.

3.4 Model Selection

To perform, model selection the data should be processed accordingly before models are created in order that models can pick the patterns faster. Particularly, categorical data were one-hot-encoded, whilst numerical values were standardized. The dataset has 16 characteristics after processing.

3.4.1 Linear Regression

Linear Regression is an interpretable model which implies a linear relationship among the dependent variable and independent variable [4]. Linear regression model performs the task to predict the value of b(dependent variable) and a (independent variable). The equation of the linear regression with b as dependent variable and a as independent variable is as shown in Eq. 3:

$$b = x + y * a \tag{3}$$

where x = Constant, y = Regression constant, b = Estimated score of the dependent variable, a = Independent variable score.

To perform linear regression in Python, we use the scikit-learn library, which provides a comprehensive set of tools for machine learning tasks. The linear regression scatter plot is as shown in Fig. 4 where Area is represented by x axis and price is represented by y axis.

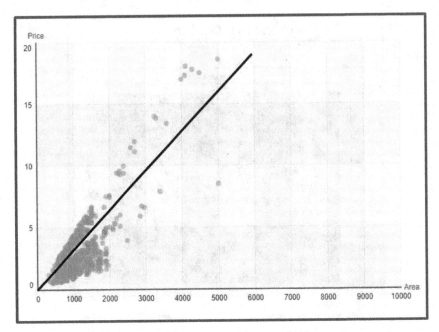

Fig. 4. Linear Regression Scatter Plot

3.4.2 Random Forest

This is a strong algorithm for ensemble learning commonly used for housing price prediction and various other regression tasks. It integrates the predictions of various decision trees to make a single prediction, resulting in more accurate and robust predictions. Some benefits of using random forests for housing price prediction includes [5]:

1. Handling Complex Relationships
2. Overfitting Reduction
3. Robustness to outliers and Noise
4. Feature Importance
5. Handling Missing Values

3.4.3 Gradient Boost

This is another popular ensemble learning technique for regression tasks like predicting house price. It works by integrating various weak prediction models, typically decision trees, in a sequential manner to improve the overall prediction accuracy. Some benefits of using Gradient Boosting for housing price prediction includes [6, 7]:

1. Ensemble of Weak Models
2. Sequential Refinement
3. Minimizing Loss Function
4. Gradient Descent
5. Feature Importance

Popular implementations of Gradient Boosting include Gradient Boosting Regression Trees (GBRT) and XGBoost [8]. These libraries offer highly optimized and efficient algorithms for training Gradient Boosting models.

3.4.4 Ensemble Model

In the proposed model we create an ensemble model for predicting house prices that combines a Gradient Boosting Regressor (traditional machine learning) and a Feedforward Neural Network (deep learning). The following are the steps used to create an ensemble model:

1. **Data Preparation:** Load and preprocess your housing price dataset.
2. **Build Individual Models:** Create a Gradient Boosting Regressor and a Feedforward Neural Network. Here, we built two individual models: a Gradient Boosting Regressor and a Feedforward Neural Network.
   i) **Gradient Boosting Regressor (GBR):** A machine learning approach called gradient boosting constructs a sequential ensemble of weak learners, typically decision trees. By concentrating on the samples that were incorrectly predicted, it seeks to rectify the mistakes of the earlier models.
   ii) **Feedforward Neural Network (NN):** One kind of deep learning model is the feedforward neural network. It is composed of linked neurons arranged in input, hidden, and output layers. Each neuron applies a weighted sum of its inputs, processes it using an activation function (like ReLU), and then sends the result to the layer below. Through training, the network has the ability to translate input features into output values.

Architecture of Feed Forward Neural Network:

Figure 5 depicts the feed forward neural network's architecture for the chosen dataset:

1. Input Layer: Number of Nodes:17
2. Hidden Layers:
   a. The ReLU activation is used by the 64 neurons in the first hidden layer.
   b. The ReLU activation is used by 32 neurons in the second hidden layer.
3. Output Layer: The Output Layer has 1 node for predicting the housing price.
4. **Training and Tuning:** Train each individual model and tune their hyperparameters. The GBR is trained by iteratively fitting decision trees to the residuals of the previous trees. Hyperparameters like the number of trees and learning rate are important to tune for optimal performance. The NN is trained by iteratively adjusting the weights of the neurons to minimize the mean squared error loss. The number of neurons, layers, activation functions, and learning rate are key hyperparameters to tune.
5. **Ensemble Model:** Combine the predictions of the individual models. After training the individual models, we make predictions using both the GBR and the NN on testing data. In ensemble model, we combine these predictions using simple averaging. This averaging technique is a basic way to combine predictions, and it can help mitigate biases from each individual model.

6. **Evaluation:** Analyse the performance of the ensemble model. We use the mean squared error (MSE) statistic to assess the ensemble model's performance. The average squared difference between the expected and actual values is measured by MSE. Greater prediction accuracy is indicated by lower MSE values.

For this, we'll use Python and the scikit-learn library for the Gradient.

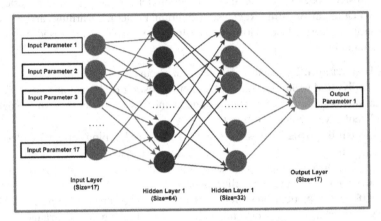

Fig.5. Architecture of Feed Forward Neural Network

Boosting Regressor and TensorFlow/Keras for the neural network.

4   Results and Discussions

To Analyse the performance of the ensemble model we use the mean squared error (MSE). The average squared difference between the expected and actual values is measured by MSE. Greater prediction accuracy is indicated by lower MSE values. The formula for calculating MSE is shown in Eq. 4:

$$MSE = n1 \sum_{i=1}^{n}(yi - y^{\wedge}i)2 \tag{4}$$

where, n is the number of data points (observations) in your dataset.

yi represents the actual housing price for the ith data point.
yi ^ represents the predicted housing price for the ith data.

R-squared (R2), also known as the coefficient of determination, is a statistical metric that expresses how much of the variance in the dependent variable (target) in a regression model is explained by the independent variables (predictors). Equation 5 displays the formula to compute R-squared:

$$R^2 = 1 - SSres \big/ SStotal \tag{5}$$

where, SSres is the sum of squared errors) between the predicted values and the actual values.

SStotal is the total sum of squares, which represents the total variance of the actual values around the mean.

An additional frequently used metric to assess a regression model's effectiveness, particularly the ability to predict home prices, is the Mean Absolute Error (MAE). The average absolute discrepancies between the values that were anticipated and the actual values are measured. MAE offers a simple method for understanding the average deviation between your forecasts and the actual numbers. Equation 6 provides the formula for determining MAE.

$$MAE = 1/n \sum_{i=1}^{n} | yi - y\char`\^i | \qquad (6)$$

where, n is the number of observations in your dataset.

yi represents the actual housing price for the ith data point.
yi ^ represents the predicted housing price for the ith data.

The performance metrics of the models discussed above are given in Table 3:

Table 3. Performance Metrics

Metric/Model	Linear Regression	Random Forest	Gradient Boost XGBoost	Ensemble Model (GBR + NN)
Mean Squared Error	703.14	358.56	159.48	**105.44**
R-squared	0.999	0.999	0.999	**0.999**
Mean Absolute Error	639.97	587.56	317.56	**26.68**

Inferences:
The following are some of the inferences from Fig. 6:

Mean Squared Error (MSE): The Linear Regression model has the highest MSE, indicating that its predictions have a higher average squared difference from the values compared to the other models. The Random Forest model has lower MSE compared to Linear Regression, indicating better predictive accuracy. The Gradient Boost model has a lower MSE compared to both Linear Regression and Random Forest, suggesting even better predictive accuracy. The XGBoost model has the lowest MSE among the Linear regression models, Random Forest model indicating the smallest average squared difference between predicted and actual values. The ensemble model (GBR + NN) is much accurate as it has least MSE values when compared to the other models, which is a good sign for the ensemble model.

R-squared: All models have extremely high R-squared values, indicating that they explain the variance in the target variable very well. This suggests that the models capture the patterns in the data effectively. Among all models the GBR+NN model shows high accuracy.

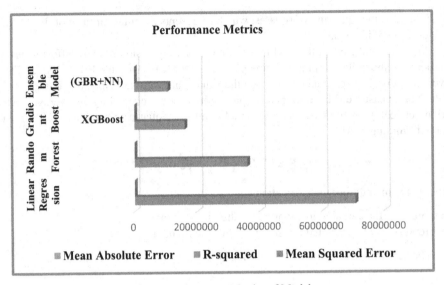

Fig. 6. Performance Metrics of Models

Mean Absolute Error (MAE): The Linear Regression model has the highest MAE, indicating that its predictions have the highest average absolute difference from the actual values. The Random Forest, Gradient Boost, and XGBoost models have lower MAE values when compared to linear regression, suggesting that the predictions are closer to actual values. The ensemble (GBR + NN) model has lower MAE value than other models which interprets that it is an accurate model.

In summary, the ensemble model (GBR + NN) performs quite well, when compared to others models Linear Regression, Random Forest, XGBoost. The choice between the ensemble model and other models might also depend on other factors such as interpretability, computational resources, and ease of deployment.

5   Conclusion

In conclusion, this paper presents a comprehensive investigation into housing price prediction in the context of smart cities. By evaluating a range of individual machine learning models and introducing a novel ensemble approach, the study contributes to advancing the accuracy and reliability of housing price prediction methods. The research assesses the effectiveness of prominent models including Linear Regression, Random Forest, Gradient Boost, and XGBoost. Additionally, it introduces a pioneering ensemble model that combines Gradient Boosting with a Feedforward Neural Network. Leveraging a diverse dataset of housing attributes, the study employs rigorous preprocessing and feature engineering techniques to enhance predictive accuracy.

Important metrics including Mean Squared Error (MSE), R-squared, and Mean Absolute Error (MAE) are used in this work to assess how well these models perform. These measures offer information on the precision of the models and their ability to account for fluctuations in home values. The study's results demonstrate that the ensemble model achieves predictive accuracy comparable to that of XGBoost, Random Forest and Linear Regression models.

References

1. Reddy, K.S., Naga Santhosh Kumar, C.H.: Effective data analytics on opinion mining, Int. J. Innov. Technol. Exploring Eng. 8(10), 2073–2078 (2019)
2. Zulkifley, N.H., Rahman, S.A., Ubaidullah, N.H., Ibrahim, I.: House price prediction using a machine learning model: a survey of literature. Int. J. Mod. Educ. Comput. Sci. 12, 46–54 (2020). https://doi.org/10.5815/ijmecs.2020.06.04
3. Santhosh Kumar, C.N., Pavan Kumar, V., Reddy, K.S.: Similarity matching of pairs of text using CACT algorithm. Int. J. Eng. Adv. Technol. 8(6), 2296–2298 (2019)
4. Agarwal, U., Gupta, S., Goyal, M.: House price prediction using linear regression in Ml (2022). https://doi.org/10.13140/RG.2.2.11175.62887
5. Adetunji, A.B., Akande, O.N., Ajala, F.A., Oyewo, O., Akande, Y.F., Oluwadara, G.: House price prediction using random forest machine learning technique. Procedia Comput. Sci. 199, 806–813 (2022)
6. Belsare, H.S., Warkar, K.V.: A novel model for house price prediction with machine learning techniques. Int. J. Sci. Res. Sci. Technol. 10(3), 743–754 (2023)
7. Zaki, J.F.W., et al.: House price prediction using hedonic pricing model and machine learning techniques. Concurrency Comput. Pract. Exp. 34, e7342 (2022)
8. Noh, B., Youm, C., Goh, E., et al.: XGBoost based machine learning approach to predict the risk of fall in older adults using gait outcomes. Sci. Rep. 11, 12183 (2021)
9. Phan, T.D.: Housing price prediction using machine learning algorithms: the case of melbourne City, Australia. In: 2018 International Conference on Machine Learning and Data Engineering (iCMLDE), Sydney, NSW, Australia, pp. 35–42 (2018). https://doi.org/10.1109/iCMLDE.2018.00017
10. Reddy, A.M., et al.: An efficient multilevel thresholding scheme for heart image segmentation using a hybrid generalized adversarial network. J. Sens. 2022, 1–11 (2022)
11. Chen, T., Carlos, G.: Xgboost: a scalable tree boosting system. In: Proceedings of the 22nd ACM SIGKDD International Conference on Knowledge Discovery and Data Mining (2016)
12. Chou, J.-S., Fleshman, D.-B., Truong, D.-N.: Comparison of machine learning models to provide preliminary forecasts of real estate prices. J. Housing Built Environ. 37(4), 2079–2114 (2022)
13. Ho, W.K.O., Tang, B.-S., Wong, S.W.: Predicting property prices with machine learning algorithms. J. Prop. Res. 38(1), 48–70 (2021). https://doi.org/10.1080/09599916.2020.1832558
14. Prakash, H., Kanaujia, K., Juneja, S.: Using machine learning to predict housing prices. In: 2023 International Conference on Artificial Intelligence and Smart Communication (AISC), Greater Noida, India, pp. 1353–1357 (2023). https://doi.org/10.1109/AISC56616.2023.10085264
15. Xiao, Y.: Big Data for comprehensive analysis of real estate market (2022). Electronic Theses, Projects, and Dissertations. 1596 https://scholarworks.lib.csusb.edu/etd/1596
16. Grybauskas, A., Pilinkienė, V., Stundžienė, A.: Predictive analytics using big data for the real estate market during the COVID-19 pandemic. J Big Data. 8(1), 105 (2021). https://doi.org/10.1186/s40537-021-00476-0. Epub 2021 Aug 3. PMID: 34367876; PMCID: PMC8329615.Li, K., & Yang, G. (2021)

17. Wang, P.Y., Chen, C.-T., Su, J.-W., Wang, T.-Y., Huang, S.-H.: Deep learning model for house price prediction using heterogeneous data analysis along with joint self-attention mechanism. IEEE Access **9**, 55244–55259 (2021). https://doi.org/10.1109/ACCESS.2021.3071306

18. Thamarai, M., Malarvizhi, S.: House price prediction modeling using machine learning. Int. J. Inf. Eng. Electron. Bus. **12**, 15–20 (2020). https://doi.org/10.5815/ijieeb.2020.02.03

19. https://www.kaggle.com/datasets/mssmartypants/paris-housing-classification

20. Manasa, J., Gupta, R., Narahari, N.S.: Machine learning based predicting house prices using regression techniques. In: 2020 2nd International Conference on Innovative Mechanisms for Industry Applications (ICIMIA), Bangalore, India, pp. 624–630 (2020). https://doi.org/10.1109/ICIMIA48430.2020.9074952

21. Madhuri, C.R., Anuradha, G., Pujitha, M.V.: House price prediction using regression techniques: a comparative study. In: 2019 International Conference on Smart Structures and Systems (ICSSS), Chennai, India, pp. 1–5 (2019). https://doi.org/10.1109/ICSSS.2019.8882834

Detection of Leaf Blight Disease in Sorghum Using Convolutional Neural Network

A Senthil Kumar[1]([envelope]), Selvaraj Kesavan[2], Kumar Neeraj[3], N Sharath Babu[3], K Sasikala[4], and Bethelegem Addisu[5]

[1] CS Cluster, Dayananda Sagar University, Bengaluru, India
angusen@gmail.com
[2] DXC Technology, Bengaluru, India
[3] Anurag University, Majarguda, India
[4] Vels Institute of Science Technology and Advanced Studies, Chennai, India
[5] Dilla University, Dila, Ethiopia

Abstract. The condition known as leaf blight significantly affects sorghum, and if it isn't treated right away, it can negatively impact the country's economy and production. This disease affects Arbaminch Gamo Gofa and other humid regions of Ethiopia where sorghum is grown. Sorghum leaf blight is typically detected through physical inspection and chemical analysis. These methods, nevertheless, are ineffective, expensive, time-consuming, and require a specialist in the field. The newly developed model, which is based on the AlexNet pre-trained model, is utilized to identify the leaf blight disease in sorghum. The main goal of this study is to detect sorghum leaf blight disease, which affects the sorghum plant's leaves. In the Gamo Gofa zone, which is where more sorghum is produced, the first 2000 original digital photos were collected. Following that, noise reduction and image reconstruction techniques were used to provide a picture for further study. Convolution, pooling, flattening, and full connection came next. The study's models were created using Keras and Tensorflow. Our test results demonstrate that the identification of leaf blight in sorghum using a convolutional neural network model is 97% accurate overall.

Keywords: Leaf Blight Disease · Sorghum · CNN · AlexNet · Gedeo Zone

1 Introduction

Agriculture is the most important sector in the Ethiopian economy, forming the foundation for the development of the entire nation. As a study demonstrates, livestock production accounts for 27% of the sector's outputs, while other areas contribute 13% of the total agricultural value added. Crop production makes up the remaining 60%. The majority of Ethiopia's small farms continue to raise grains primarily for home use as well as export. Nearly three-quarters of the total land under cultivation is used to grow cereals [1].

One of the most popular cereal crops in Ethiopia is sorghum. Millions of underprivileged Ethiopians rely on it as their primary source of food in order to survive. Ethiopian

© The Author(s), under exclusive license to Springer Nature Switzerland AG 2024
S. Satheeskumaran et al. (Eds.): ICICSD 2023, CCIS 2122, pp. 123–134, 2024.
https://doi.org/10.1007/978-3-031-61298-5_10

farmers make excellent use of every part of this plant. Sorghum is a crop that thrives in a variety of agro-ecologies, especially those that are moisture-stressed and where food insecurity is rife [2]. Many different diseases can harm sorghum, which can result in significant output and financial losses. These illnesses include head smut, leaf spot, leaf rust, leaf blight, and sooty stripe. Few of these diseases are widespread and do not pose a serious threat to any one industry, yet their ubiquity may result in a substantial net loss overall. Some diseases are easy to spot because they have recognisable signs that notably lower yields. Numerous pathogenic species, including fungi, bacteria, and viruses, are responsible for sorghum diseases. These could be addressed with the customized deep learning model 'MaizeNet' for disease detection, severity prediction, and crop loss estimation [3] and also plant disease could be classified with de facto paradigm of crop disease categorization [4]. The model reports the highest accuracy of 98.50%. Also, the authors perform the feature visualization using the Grad-CAMFor a number of reasons, some diseases do not exist in a particular region. Exserohilum turcicum is the source of the foliar disease known as sorghum leaf blight. This disease is found in many humid regions where sorghum is cultivated. These regions have a moderate temperature range of 15.5 to 26.7 °C. [4]. This sorghum leaf blight cannot establish itself in dry weather. Small reddish or tan spots may appear on the leaf of infected seedlings. These spots become bigger, the leaf wilts and turns purplish grey, and the seedling may finally die as the illness progresses. Long elliptical lesions that may be reddish-purple or yellowish tan grow on older leaves in more mature plants. Depending on varied degrees of resistance, these lesions differ in size and colour. On older leaves, the majority of the lesions start, then spread to younger leaves. On elder plants, lesions have reddish edges and centres that range from yellow to grey. Up to 50% of yield losses are likely if the disease is detected before boot stage [5].

A mechanism for identifying plant diseases is the main obstacle in sorgum [6]. In the Arbaminch Gamo Zone, the disease leaf blight has a serious impact on sorghum. This illness is widespread in Ethiopian regions where sorghum is produced. Unless it is handled at an early stage, leaf blight in sorghum plants is a severe disease. Small reddish or tan patches appear on the leaves of diseased seedlings. These spots become larger as the illness worsens and the seedling eventually perishes [4]. Physical observation and laboratory testing are the conventional methods for diagnosing the sorghum leaf blight. However, the conventional methods of sorghum leaf blight detection have their own limitations. These processes could be inaccurate, laborious, unpredictable, dependent on knowledge of plant diseases, ineffective, and requiring specialised lab equipment. Microorganisms are required for the survival and growth of laboratory experiments used to diagnose diseases. Microorganisms need to be able to survive and thrive, and a good growing environment is essential. Nutrient solutions called culture medium are used in labs to cultivate microorganisms. Microorganisms take 5 to 7 days to grow [8]. Because of this, identifying the mechanism of sorghum leaf blight illness takes time. As a result, the disease significantly contributes to the decline in sorghum crop production and profitability. The goal of this work is to use CNN to construct an automated system for identifying the sorghum disease leaf blight.

2 Research Design and Model Development

This study focuses on identifying and categorizing sorghum leaf diseases according to their symptoms as they manifest on the plant's leaves. Sorghum's leaf blight manifests physically as a variety of changes to the leaves' morphology, forms, and hues. Sorghum leaf disease comes in a variety of forms, which are covered in chapter two. This section will show how sorghum leaf blight is detected and classified using machine learning approaches.

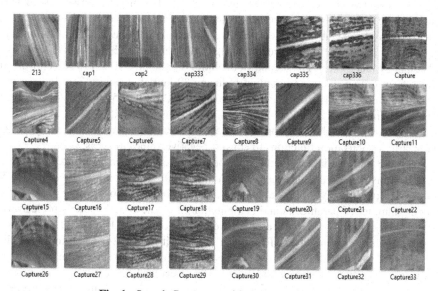

Fig. 1. Sample Preprocessed Sorghum Leaf Image

To deal with the above challenges, the following process model is used as depicted in Fig. 1 that shows that the process consists two main phases: Training phase and Testing phase. In the following, detail of each phase is described. Diseased and healthy leaves of sorghum were collected from Arbaminch Chano AARC. Total 2000 infected and healthy leaves of sorghum were taken from AARC. These have three channels Red(R), Green (G) and Blue(B). Before the model is trained, the pictures in the dataset for the deep CNNs classifier are preprocessed. The normalization of picture size and format is one of the most important activities. To preprocess an image for this study, it is resized to 224*224 pixels and, if necessary, made grayscale. The stage of testing is crucial for evaluating how well the best trained model performed. The model that has been trained is given a set of query photos to be evaluated. The trained model is used to check whether the leaf is impacted or not after performing feature collection and pre-processing, although the datasets are brand-new.

2.1 Proposed Model Development

The convolutional neural network (CNN) is used in this study to pinpoint the ailment known as sorghum leaf blight. An input layer, an output layer, and multiple hidden layers are all present in a CNN, a machine learning approach. Figure 2 shows the proposed model with all the layers involved and detailed description.

As shown in Fig. 2 it is the first layer in which the image can be given as input directly to the CNN model. But here data augmentation is done before passing the images to the CNN model. The input layer of the proposed CNN model accepts RGB images of size 224×224 with two classes (normal leaf and affected leaf). This layer passes raw images as input to the first convolution layer without any computation or change. Every image is to be considered as a matrix of pixel values. As shown in Fig. 3 the first Convolution(C1) filters $222 \times 222 \times 3$ input images by using 32 kernel size with $3 \times 3 \times 3$ filter size and default stride(1pixel). The second Convolution(C2) filters $110 \times 110 \times 3$ convolved images from output of C1 and pooling by using 32 kernel size with $3 \times 3 \times 3$ filter size and default stride (1pixel).The third Convolution(C3) filters $54 \times 54 \times 3$ convolved images from output of C2 and pooling by using 32 kernel size with $3 \times 3 \times 3$ filter size and default stride(1pixel). The proposed model's convolutional layers all employ ReLU nonlinearity as activation functions to convert negative values to zeros.

Since the data learned by ConvNet would not be linear, an additional operation called ReLu is conducted after each convolution to introduce nonlinearity to the CNN model. Max pooling layer plays a great role in controlling the overfitting problem by reducing number of parameters and computation in the network. There is no learning process in this layer. This layer is used after each convolution (C1, C2 and C3). At each Convolution, maxpooling layer of filter size 2×2 with stride 2 is used. In this study, FC1 and FC2 layers with probability 0.5 has been utilized to overcome the problem of overfitting. The input of the sigmoid classifier is a vector of the features yielding from the learning process and the output is a probability that an image belongs to a given class. In this study two fully connected layers are used. The first FC1 is with 64 dense and ReLU activation function. The last FC2 is with dense of 2 (number of class labels) and Sigmoid activation function in order compute the classes scores. In the course of training, the CNN learns what kinds of features to extract from the input images. Processes like convolution and pooling are employed during feature extraction. Convolutional layers of CNN use feature extraction as their primary function, which is how this layer works. The image has a $224 \times 224 \times 3$ pixel size at the input, indicating that it has 224 pixels in both height and width and that it includes 3 channels, generally referred to as RGB (red, green, and blue). Different matrix values are present in each pixel channel. The chosen filter value will be convolved onto the input. The final component of the suggested CNN model is the classifier. It is merely a dense layer, often known as an artificial neural network (ANN). This portion of the CNN model, where data dimensions are modified such that data may be categorised linearly, is also known as a fully **connected** process.

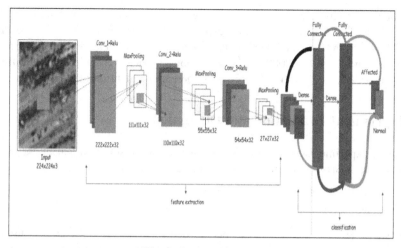

Fig. 2. Proposed Model

3   Research Methodology

3.1   Dataset

The images were collected from Chano Woreda. It is found in Arbamich Gamo Zone. Arbaminch Agricultural Research Center has experiment station from Chano Woreda. The Center conducts different experiments related to plants in this woreda. The images were taken from sample of sorghum planted for the purpose of identifying diseases appeared in Chano Woreda.Chano Woreda Experiment station is selected to get high quality images and it is close to AMU. The images taken from farming land may not be good in quality due to different reasons. But here in experiment station, the treatments given to each plant is very high and we could get good growing sorghum leaf at flowering stage. The photographs were taken after 11:00 local time throughout the day to avoid being distracted by sunlight.

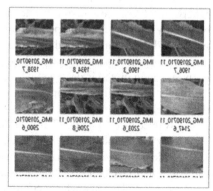

Fig. 3. Sample collected images of both Diseased and Normal Soghum leaf

Photos of sorghum leaves are healthy sorghum leaves and sick sorghum leaves that have bacterial leaf blight. Agricultural experts were consulted to ensure the accuracy of the classes in the collected dataset. The photos were shot at a 150 mm distance and at a resolution of 4160 × 2080 pixels. The photos, which were initially at a different resolution, are resized to 224 × 224 pixels to prepare the dataset for training.

4   Experiment Evaluation and Discussion

A series of experiments utilizing an actual data set of the sorghum leaf blight disease are carried out to assess the efficacy of the suggested model.

4.1   Dataset

In this study, 80% of the dataset has been utiled for training, while 20% is used for validation. There are several models examined, each with a different architecture and learning rate. The photos in our collection can be divided into two groups. Sorghum leaves that are healthy and those that are sick fall into these two groups. 1000 images of sorghum leaves in good health and 1000 images of leaves with disease were taken from the AARC Chano Subcenter. 2000 total original photographs were utilized for the experiment, but 12000 more images were used instead. But just training sample augmentation is carried out. Aiming to lessen overfitting during training, augmentation is used for training samples. Trial and error was used to choose the network's parameters, including the learning parameter and filter size. Additional versions are produced by Rotation, Shear range, Horizontal flip, Room_range, and vertical_flip after the initialization of the original pictures.

4.2   Training Model

As per Sabbir Ahmed [40], For a suggested CNN-based system, picking the right architecture can be a difficult problem because it takes a lot of fiddling to get the hyperparameters just right. It can be challenging to determine the ideal layer count, filter size, stride, padding, and other parameters. There is no unambiguous standard established by researchers, and these are not simple questions. This is due to the fact that a network will be heavily reliant on data type that varies with image size, data complexity, the computing hardware available, and much more.

By examining the dataset presented above and testing with various hyper parameters, the best performing model is designed with following detail (Fig. 4).

The suggested model needs input photos with a 224 × 224(RGB) resolution. The input layer receives the leaf picture, which has the dimensions 224 × 224 × 3. The hidden layers are traversed by the photos. Convolutional, Rectified Linear Unit, and Maxpooling layers make up each hidden layer. The Each layer comprise 32, 32, 32 filters with a 3 × 3 size, stride 1, and no padding by default (p = 0). The ReLU layer that follows each convolutional layer improves the training process and network performance. For all three maxpooling layers, the layer configuration is 2 × 2 with stride 2 and padding 0. In this model, the required number of classes for specifying the probability distribution

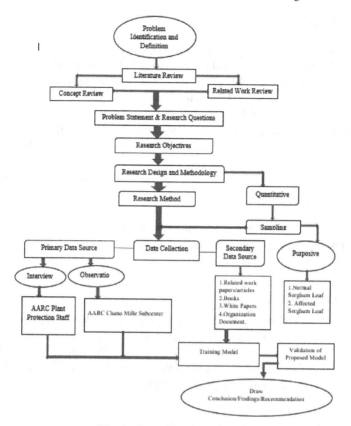

Fig. 4. Research Flow Diagram

is decided based on the second fully linked layer. The aforementioned CNN model is trained using the SGD (Steepest Gradient Descent) technique. The layer implementation of the suggested model is shown in the following Table 1.

Table 1. Architecture of the Proposed Model

Layer	Filter size	Output size
Input		224 × 224
Convolution layer 1	3 × 3	222 × 222 × 32
Maxpooling layer 1	2 × 2	111 × 111 × 32
Convolution layer 2	3 × 3	110 × 110 × 32
Maxpooling layer 2	2 × 2	55 × 55 × 32
Convolution layer 3	3 × 3	54 × 54 × 32
Maxpooling layer 3	2 × 2	27 × 27 × 32

4.3 Hyper Parameters

Our model uses the stochastic gradient descent (SGD) algorithm to discover the optimal biases and weights for the neural network to minimize the loss function. The SGD algorithm selects a limited number of training inputs at random to carry out its learning process. This batch size is set at 16 and the learning of 0.001 is chosen in this model. It refers to how quickly a function traverses the search space. Although it takes longer training time, a slow learning rate produces outcomes that are more precise. The momentum also affects how soon the SGD algorithm converges on the ideal location. The value is 0.9.

4.4 Experimental Results

By altering the parameter values, several experiments were run. The ensuing tables display the findings of each experiment. An experiment was run to determine the impact of dropout, learning rates, and filter size.

4.4.1 Effect of Learning Rate

The learning rate was found to have the most impact in this experiment part. The learning rate controls how quickly the gradient descent method finds the best answer. Learning rate (0.01, 0.001, and 0.0001) was changed from the original learning rate. From the results obtained in Table 2, ii is observed that the learning rate of 0.001 provides superior performance with the validation accuracy of 0.98.

Table 2. Accuracy and loss analysis for various learning rates

Learning rate	Training accuracy	Validation accuracy	Training loss	Validation loss
0.001	0.89	0.86	0.23	0.27
0.001	0.97	0.95	0.16	0.18
0.0001	0.91	0.90	0.21	0.24

Dropout regularization is considered in this experiment to address the issue of over fitting dropout regularization. The suggested model's performance can differ depending on whether it was trained with or without dropout regularization, as shown in table 5.5, where the highest performance was obtained with a probability value of 0.5.

4.4.2 Effect of Numbers of Epochs

This hyperparameter determines how many times the learning algorithm will traverse the whole training datasets. Every sample in the training dataset has had a chance to update the internal model parameters after one epoch. In this model, 1600 training samples are utilized with the batch size of 16. Total number of 120 epochs are performed in this study.

The dataset is therefore split into 100 batches, each containing 16 samples. Each time 16 samples were taken, the model weights would have been modified. The model was updated or 100 batches were used in one epoch, according to this. The entire dataset is traversed by the model 120 times. Throughout the entire training procedure, there were 12000 batches in all. As the number of epochs rises, accuracy, according to experimental findings.

```
Epoch 112/120
100/100 [==============================] - 326s 3s/step - loss: 0.0755 - acc: 0.9737 - val_loss:
0.0684 - val_acc: 0.9775
Epoch 113/120
100/100 [==============================] - 328s 3s/step - loss: 0.0885 - acc: 0.9669 - val_loss:
0.0667 - val_acc: 0.9800
Epoch 114/120
100/100 [==============================] - 342s 3s/step - loss: 0.1067 - acc: 0.9569 - val_loss:
0.0736 - val_acc: 0.9750
Epoch 115/120
100/100 [==============================] - 341s 3s/step - loss: 0.0695 - acc: 0.9750 - val_loss:
0.0833 - val_acc: 0.9750
Epoch 116/120
100/100 [==============================] - 332s 3s/step - loss: 0.0945 - acc: 0.9644 - val_loss:
0.0822 - val_acc: 0.9700
Epoch 117/120
100/100 [==============================] - 358s 4s/step - loss: 0.0825 - acc: 0.9662 - val_loss:
0.1075 - val_acc: 0.9600
Epoch 118/120
100/100 [==============================] - 343s 3s/step - loss: 0.0839 - acc: 0.9694 - val_loss:
0.0736 - val_acc: 0.9750
Epoch 119/120
100/100 [==============================] - 379s 4s/step - loss: 0.0798 - acc: 0.9725 - val_loss:
0.0574 - val_acc: 0.9825
Epoch 120/120
100/100 [==============================] - 378s 4s/step - loss: 0.0861 - acc: 0.9706 - val_loss:
0.0670 - val_acc: 0.9750
```

Fig. 5. Training Model Progress

The summary of the effects of the epochs when 30, 60, and 120 are presented in Table 3 below. The results of the experiment indicate that the best performance was obtained with a 120 epoch. The proposed model achieves the best validation accuracy of 0.97 for 120 epochs.

Table 3. Performance analysis for number of epochs

Epochs	Training accuracy	Validation accuracy	Training loss	Validation loss
30	0.93	0.95	0.16	0.10
60	0.95	0.97	0.12	0.09
120	**0.97**	**0.98**	**0.07**	**0.05**

The model performs better on training data as well as validation data, as seen in Fig. 5. Validation accuracy is somewhat higher than training accuracy. A little over fitting is typical, and methods such as dropout were used to control higher over fitting levels (Fig. 6).

Figure 7 illustrates how a model's accuracy and loss during training for validation data may vary depending on the circumstance. Loss ought to be decreasing with every passing epoch. While validation accuracy and training accuracy are improving, validation loss and training loss start to decline.

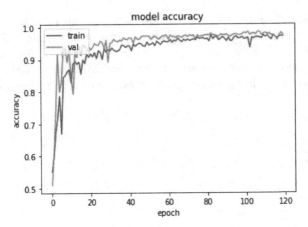

Fig. 6. Training Accuracy vs Validation Accuracy

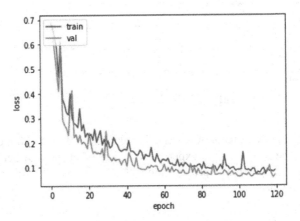

Fig. 7. Training Loss vs Validation Loss

5   Conclusion

One of the most prevalent diseases to affect sorghum leaves in the Arbaminch Gamo zone is sorghum leaf blight. Plant protection experts employed traditional methods, such as observation and laboratory tests, to find the source of the sorghum leaf blight disease. Since these methods are time-consuming and expensive, it is necessary to develop a better model. The development of a Convolutional Neural Network-based disease detection model for sorghum leaf blight aids in the disease's accurate identification. The Keras library was used to implement the proposed model, and Tensorflow was supported. Arbaminch Agricultural Research Center's (chano subcenter) images of healthy and damaged sorghum leaves were gathered, and after that, image preprocessing and image augmentation were used. Hyperparameter values were changed to undertake various experiments. Affected sorghum leaves can be distinguished from healthy ones using this model. The Keras library and Tensorflow were used to implement the proposed model. Arbaminch Agricultural Research Center's (chano subcenter) photos of healthy and damaged sorghum leaf images were gathered, and after that, image preprocessing and image augmentation were used. Hyperparameter values were changed to undertake various experiments. For identifying and classifying the sorghum leaf blight disease, this result is good. It is advised to conduct additional study on identifying various sorghum leaf diseases using other deep learning frameworks for model training and testing.

References

1. Gebre-Selassie, D.A.: A review of Ethiopian agriculture: roles, policy and small-scale farming, koperazzijoni internazzjonali-malta, p. 10 (2010)
2. P. D. S. A. Alemayehu Seyoum Taffesse, Crop Production in Ethiopia: Regional Patterns and Trends. In: International Food Policy Research Institute, March 2011
3. Kundu, N., Rani, G.: Disease detection, severity prediction, and crop loss estimation in MaizeCrop using deep learning. Artif. Intell. Agric. **6**, 276–291 (2022). ISSN 2589–7217
4. Lee, S.H., Goëau, H., Bonnet, P., Joly, A.: New perspectives on plant disease characterization based on deep learning. Comput. Electron. Agric. **170**, 105220 (2020). ISSN 0168–1699
5. Rajesh, V., Naik, U.P., Mohana.: Quantum convolutional neural networks (QCNN) using deep learning for computer vision applications. In: Proceedings of the 2021 International Conference on Recent Trends on Electronics, Information, Communication & Technology (RTEICT), pp. 728–734, Bangalore, India, August 2021
6. Chen, S., Zhang, K., Zhao, Y., et al.: An approach for rice bacterial leaf streak disease segmentation and disease severity estimation. Agriculture **11**(5), 420 (2021)
7. Poornappriya, T.S., Gopinath, R.: Rice plant disease identification using artificial intelligence approaches. Int. J. Electr. Eng. Technol. **11**(10), 392–402 (2022)
8. Khan, R.U., Khan, K., Albattah, W., Qamar, A.M.: Image-based detection of plant diseases: from classical machine learning to deep learning journey. Wirel. Commun. Mob. Comput. **2021**, 13 (2021). Article ID 5541859
9. HE Dun-chun1, Z. J.-S. X. L.-H.: Problems, challenges and future of plant disease management: from an Ecological point of view, Elsevier, pp. 705–715 (2016)
10. SRA, O.J.: Pioneer (2019)
11. Davies, E.: Computer Vision: Principles, Algorithms, Application Learning, Mara Conner, United Kingdom (2018)
12. Sullivan, W.: Machine learning: beginners Guide Algorithms (2017)
13. Salman Khan, H.R.A.A.S.B.: A Guide to Convolutional Neural Networks for Computer Vision, Morgan & Claypool, Southern California (2018)
14. II, T.B.: Introduction to Deep Learning Using R, San Francisco, California, Taweh Beysolow II, USA (2017)
15. Hamed Habibi, J.: Guide to Convolutional Neural Networks, Springer, Spain (2017)
16. Kim, P.: MATLAB Deep Learning With Machine Learning, Neural Networks and Artificial Intelligence, Phil Kim, Korea (2017)
17. Mohit Sewak, M.R.K.P.: Practical Convolutional Neural Network, Packt Publishing, Birmingham (2018)
18. Aleshin-Guendel, S.: Examining the Structure of Convolutional Neural Network," Computer Science Honors Thesis Boston College, Boston (2017)
19. Torrez, J.: First Contact with Deep Learning, UPC Barcelona Tech: Kindle (2018)
20. Sahla, N.E.: A Deep Learning Prediction Model for Object Classification, College of Mechanical, Maritime and Materials Engineering (3mE) · Delft University of Technology (2018)

21. Enquehone, A.: Maize leaf diseases recognition and classification based on imaging and machine learning techniques. Int. J. Innov. Res. Comput. **5** (2017)
22. Aarju Dixit, S.N.: Wheat leaf disease detection using machine learning method. Int. J. Comput. Sci. Mob. Eng. **7**(5), 124–129 (2018)
23. Arivazhagan, S.L.S.: Mango leaf diseases identification using convolutional neural network. Int. J. Pure Appl. Math. **120**, 11067–11079 (2018)
24. Chen, J., Chen, J., Zhang, D., Sun, Y., Nanehkaran, Y.A.: Using deep transfer learning for image-based plant disease identification. Comput. Electron. Agric. **173**, 105393 (2020)
25. Osinga, Deep Learning CookBook, O'Reilly Media, USA (2018). **2**(5), 99–110 (2016)
26. Author, F., Author, S.: Title of a proceedings paper. In: Editor, F., Editor, S. (eds.) Conference 2016, LNCS, vol. 9999, pp. 1–13. Springer, Heidelberg (2016)
27. Author, F., Author, S., Author, T.: Book title. 2nd edn. Publisher, Location (1999)
28. Author, F.: Contribution title. In: 9th International Proceedings on Proceedings, pp. 1–2. Publisher, Location (2010)
29. LNCS Homepage. http://www.springer.com/lncs. Accessed 21 Nov 2016

Data Security for Internet of Things (IoT) Using Lightweight Cryptography (LWC) Method

R. Siva Priya[1], V. Shunmughavel[1], B. Praveen Kumar[2]([⊠]), and E. R. Aruna[3]

[1] Department of CSE, Sethu Institue of Technology, Kariapatti, India
[2] Department of EEE, Vardhaman College of Engineering, Hyderabad, India
praveenbala038@gmail.com
[3] Department of IT, Vardhaman College of Engineering, Hyderabad, India

Abstract. The notion of Internet of Things (IoT) is significant in the continuing development of the Internet since it describes the creation of a network of tiny objects that may link countless numbers of devices from different platforms. In order to enable an efficient end-to-end communication – Ensuring total security usually requires the implementation of an algorithm with a symmetric key on the final nodes. For applications with few assets, like those run on batteries, cryptographic functionality with low energy consumption is essential. Using the lightweight symmetric key approach can result in end devices using less energy. In order to defend IoT systems against memory heap assaults, this effort aims to build a security paradigm. For this purpose, a Light Weight Cryptography (LWC) model based on Elliptic Curve Cryptography (ECC) is deployed in this work. Moreover, the one-way hash algorithm is used to protect the memory heap with high efficiency and attack detection rate. Based on the computational time metrics, the suggested LWC-ECC model's performance and outcomes are verified and compared.

Keywords: Data Security · Internet of Things (IoT) · Light Weight Cryptography (LWC) · Next Memory Address Occupation (NMOA) attacks · Elliptic Curve Cryptography (ECC) · One time hash function

1 Introduction

Many restricted devices are linked with the Internet via an emerging computing environment termed as "Internet of Things (IoT)" [1, 2]. Through the network, the devices communicate with one another and give us new insights. Security of limited end nodes is crucial for exploiting this new environment. The network could suffer severely if any of the nodes were attacked. However, due to the limited available resources, it is difficult to implement an appropriate cryptographic functionality on restricted devices. The notion of IoT is significant in the continuing development of the Internet since it describes the creation of a network of tiny objects that may link countless numbers of devices from different platforms. In practical terms, IoT devices have a finite number of security gates, limited storage, and processing power [3]. Any lightweight cipher's design goal is to

© The Author(s), under exclusive license to Springer Nature Switzerland AG 2024
S. Satheeskumaran et al. (Eds.): ICICSD 2023, CCIS 2122, pp. 135–144, 2024.
https://doi.org/10.1007/978-3-031-61298-5_11

enhance security while requiring less hardware, computational capacity, and estimation. These resource-constrained devices cannot transfer data securely using traditional cryptographic techniques. Cryptography is one of the well-known security model used for enabling a reliable data sharing and communication in networks, which is based on the mathematical principles and a system of computations known as algorithms. Also, it modifies information (cypher) in a variety of ways, making it challenging enough for someone to break it [4]. These algorithms are used to create cryptographic keys, perform digital signature and verification, and secure internet browsing as well as private communications like emails and credit card transactions. The Advanced Encryption Standard (AES), Rivest Shamir Adleman (RSA), Data Encryption Standard (DES), and Elliptic Curve Cryptography (ECC) are the most extensively used advancing cryptographic systems [5] in present times.

Moreover, the LWC [6] is still in the development stage. But in order to move forward with the data processing requirements of IoT, it requires an effective LWC solutions. This covers the prevention of potential new threats or attacks as well as ultra-high speed transmission with very low latency, availability, open source capabilities, green networking, minimal energy use, and cost. Moreover, the lightweight systems aim to have smaller key sizes, less memory usage, and faster implementation. This allows for the usage of less resources compared to substantial weighted solutions. Any lightweight method can fit in, but there are no restrictions; instead of the most important parameters like the number of blocks, algorithmic code tactics, processor cycles, and many other factors are prioritized [7]. The objective of developing a lightweight algorithm design is to compromise on an assortment of features, including the method's performance, encryption potential, and minimal resource requirements. The objectives of this paper are given below:

- A light weight cryptographic model has been deployed to protect IoT systems against security breaches.
- A new security paradigm is developed with the use of Elliptic Curve Cryptography (ECC) model with increased computational speed and minimized system complexity.
- Moreover, the one-way hash algorithm is used to protect the memory heap with high efficiency and attack detection rate.

The following sections are made up of the remaining parts of this article: The comprehensive literature overview on cryptography-based IoT security is presented in Sect. 2. The proposed LWC based ECC paradigm for IoT security is fully explained in Sect. 3. The performance outcomes of the suggested LWC model are presented in Sect. 4 utilizing various evaluation measures. The overall paper is concluded with findings and future research in Sect. 5.

2   Related Works

Among the cyber attacks of today, distributed denial of service is the most notable and significant. It was viewed as a potent assault on the current Internet community. In the context of secure remote access via near-field communication, an innovative technique is provided for safeguarding element-based cooperative verification and IoT verification with an end-user computer, including a smartphone or other handheld device [8].

An updated protocol for mutual locality authentication and attestation will put a stop to distant confidence attestation and anonymous mutual authentication between secure elements. Following such attacks, more advanced security protection techniques that try to stop the operation of injected terminal code have evolved as a more secure way to deal with malicious code [9]. An approach for controlling a program's execution by altering an operation's return address through an attack vector. Return-oriented programming codes are little chunks of code that are placed one at a time in an exploitable memory to carry out the attacker's objectives. The complex interdependencies and heterogeneities among the constituent elements of cyberphysical systems, along with their openness, error-proneness, and challenging operating conditions, open the door to increased vulnerability to malicious security attacks [10]. Numerous hurdles posed by soft errors, process variation, and temperature-induced dark semiconductor problems led to the development of numerous reduction strategies at different layers of the cyber physical system [11]. The classification model of LWC is represented in Fig. 1.

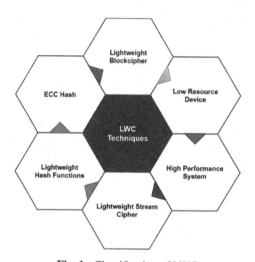

Fig. 1. Classification of LWC

3   Proposed Methodology

Security hazards [12] arise when a third party attempts to steal a session key using a guessing or brute force attack to access the matching heap address that indicates the task name. As a result, the information that was located in memory is pertinent to the task address that was now being executed. Executing undesired tasks or halting any that are already in progress will result in a resource attack. Services may suffer as a result of time wasting attacks or service denial attacks. Also, the likelihood of such risks can significantly increase, when the attacking operations are executed in the target system that burns the appropriate devices. A lack of ability to successfully manage the affected equipment or services could result in a prolonged crisis that affects the nearby devices. Therefore, in response to these hazards we offer a solution to the issue of how to avoid the NMOA, which typically enables intruders to compromise and hijack the currently active session. Moreover, a manipulated NMOA leads to exploiting garbage on the memory heap. The proposed model illustrates how the system was breached, where the hacker taking advantage of the delay between the device selection and activation of particular function. Additionally, we provide a security paradigm that protects the application's functionality from intrusion or alteration by embedding a simple garbage encryption technique that runs prior to the end of run time at the bottom of each object code file. Better identification of harmful attacks aimed at IoT-enabled environments is made possible by this security paradigm. Our suggestion was taken into consideration as an effective IoT attack mitigation, due to the demand for ongoing IoT monitoring. It ensures that no memory address is sent to an adversarial machine engaged in guessing, brute-forcing, or NMOA attacks. Garbage collection is done via encryption at the end of any item's runtime. The proposed approach makes it impossible to locate a message that generates a specific hash, not even by using an array of matched hashes or a brute force assault on all potential inputs to determine whether they match. Replay attacks cannot be used against the cryptographic hash function or the one-time key (Fig. 2).

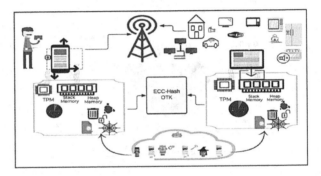

Fig. 2. Proposed framework

3.1 Light Weight Cryptography (LWC)

In this study, the major reasons of applying the LWC in IoT are listed below [13]:

- To facilitate effective communication from end to end—In most cases, symmetric key algorithms have to be applied at the end nodes in order to guarantee end-to-end security. Low energy consumption during cryptographic operations is a prerequisite for low-resource devices, including battery-powered gadgets. Through the application of the lightweight symmetric key approach, final equipment can consume less energy.
- Compatibility with devices using less energy - The lightweight cryptographic approaches have a small logical impact than the traditional cryptographic methods. The lightweight cryptographic primitives enable the usage of more network connections with devices that consume less resources.

In this paper, the learning task behavior problem is resolved with the use of LWC approach. At first, the protection process is performed to prevent the network from attacking activities. Then, the hijacking operations impersonating attacks are predicted with proper user authentication. The second stage is to use a lightweight cryptographic hash function for garbage collection in the running object to render access violations more likely to occur. The appropriate heap address is violated via a brute force attack or learning-based guessing. The task content or object thread address is the data that was obtained from the memory location. At all times, terminating unneeded tasks or quitting active threads wastes resources. Services will be continually turned on and off as a result of waste of time attacks or service denial attacks, which will break the system. The worst-case scenario is when a task is carried out and burns or damages IoT equipment. In order to provide strong security with reduce dynamic deterioration, a stream-cipher technique is developed based on reliable chaotic maps and one-time hash keys.

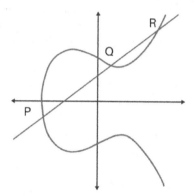

Fig. 3. Elliptic curve cryptography

3.2 Elliptic Curve Cryptography (ECC)

Typically, the ECC [14] is one of the most emerging and popular cryptographic technique widely used in many IoT security applications. Here, the purpose of using technique is to secure memory of IoT networks. It provides more privacy with a less key size than other cryptographic methods. The traditional method of mapping the characters to quadratic points in the elliptic curve has been discontinued in this study in favor of a new method. For the purposes of decrypting and encrypting communications over the internet, ECC emphasizes pairs of public and private keys. In relation to the Rivest-Shamir-Adleman (RSA) cryptographic algorithm, the ECC is frequently used for data security applications [15]. Also, it uses prime factorization to accomplish one-way encryption for items like emails, data, and software. The ECC model is shown in Fig. 3.

3.3 Security Analysis

It is becoming more and more important that IoT is actually protected from widely recognized vulnerabilities. If the level of security for IoT devices isn't constantly improved, attackers will find ways to take advantage of weaknesses to suit their own needs. Everything done in the event of an IoT system spyware will be watched by the intruder [16]. For learning how IoT devices work, the data from the currently running session and its RAM heap locations are going to be transmitted out. The user chooses the device in the network before they selects the appropriate task, as seen in Fig. 4. When the device memory heap session is attacked, the tasks that are associated with the connected device or its address in the pertinent memory are updated. Additionally, the services do not identify the address when an individual switches to the device's functionalities to select an operation or task need to be updated. An incorrect choice leads to a bad selection for its function. The attacker must initially select an operational service before the request is postponed and the device does not appear or report a conflict error before the new address is processed or ignored. In the proposed system, it is required to calculate the behavior of the programs during runtime in order to recognize several harmful attacks [17], such as ROP and heap buffer overflow. The contributors have additionally created a certification methodology with high-level-based features to manage the process of verification and the constantly changing behavior collection together. The use of machine learning techniques of various types have been used to explicitly verify the constantly changing behavior.

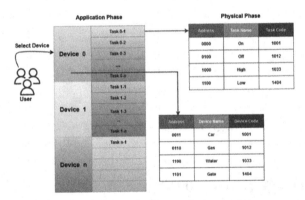

Fig. 4. Physical address mapping model

4   Results and Discussion

This section evaluates and contrasts the performance of the proposed LWC-based ECC mechanism with other encryption methods using the parameters of encryption and decryption time [18]. The encryption time is compared in Figs. 5 and 6 for various memory sizes, including 128 KB, 512 KB, 1024 KB, 2048 KB, 4096 KB, 8192 KB, and 16,384 KB, respectively. Encryption time is typically defined as the time needed to produce the cypher text. The decryption time is the same as the time needed to produce the original text. Moreover, these parameters are computed with respect to different memory heap size ranging from 2000 KB to 16000 KB as shown in Figs. 7 and 8 respectively. Overall, the results show that the LWC-ECC technique effectively reduces the encryption and decryption time consumption in the suggested model.

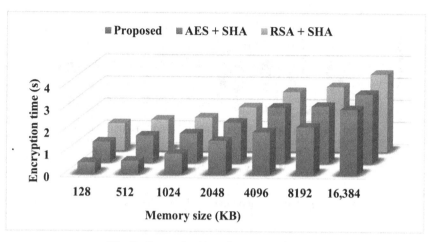

Fig. 5. Comparison based on encryption time

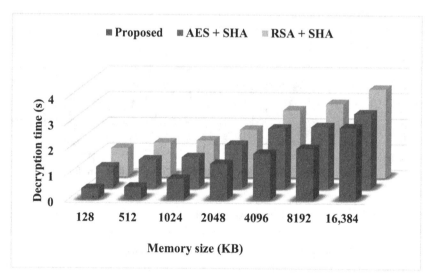

Fig. 6. Comparison based on decryption time

R. Siva Priya et al.

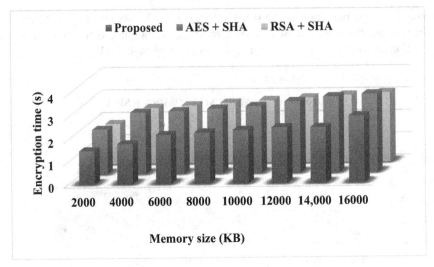

Fig. 7. Encryption time Vs Memory heap size

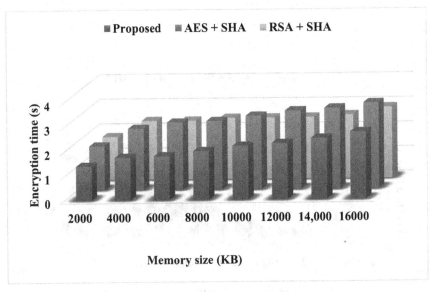

Fig. 8. Decryption time Vs Memory heap size

5 Conclusion

Typically, the operating system has the responsibility to control and monitor the running applications of memory heap. There are still vulnerabilities despite the inbuilt safeguards. In this work, a systematic method is developed to solve the garbage collection vulnerabilities with the use of LWC. Here, a well-known technique is implemented to remote acquire the physical location in a memory heap. We used a procedure with two phases to finish it. In the beginning, we deployed a TPM unit device to safeguard the CPU against any unauthorized access. We created and tested a cryptographic hash method that encrypts any garbage collection object during the second stage. To prevent IoT from NMOA that is one of the riskiest threats in the modern cyber world, the hash makes use of the one-time key algorithm, which securely encrypts trash collection. It uses a simple but incredibly effective assault tactic, which presents a massive challenge to the current Internet culture. The main components of memory heap assaults and the garbage collection cryptography system—which boosted security and thwarted malevolent external attacks—have been delineated.

References

1. Qaid, G.R.S., Ebrahim, N.S.: A lightweight cryptographic algorithm based on DNA computing for IoT devices. Secur. Commun. Netw. **2023**, 1–12 (2023). https://doi.org/10.1155/2023/9967129
2. Chauhan, J.A., Patel, A.R., Parikh, S., Modi, N.: An analysis of lightweight cryptographic algorithms for IoT-Applications. In: Advancements in Smart Computing and Information Security: First International Conference, ASCIS 2022, Rajkot, India, November 24–26, 2022, Revised Selected Papers, Part II, pp. 201–216 (2023).
3. El-Hajj, M., Mousawi, H., Fadlallah, A.: Analysis of lightweight cryptographic algorithms on IoT hardware platform. Future Internet **15**, 54 (2023)
4. Abdulraheem, M., Awotunde, J.B., Jimoh, R.G., Oladipo, I.D.: An efficient lightweight cryptographic algorithm for IoT security. In: Information and Communication Technology and Applications: Third International Conference, ICTA 2020, Minna, Nigeria, November 24–27, 2020, Revised Selected Papers 3, pp. 444–456 (2021).
5. Gunathilake, N.A., Al-Dubai, A., Buchana, W.J.: Recent advances and trends in lightweight cryptography for IoT security. In: 2020 16th International Conference on Network and Service Management (CNSM), pp. 1–5 (2020)
6. Fan, R., Pan, J., Huang, S.: ARM-AFL: coverage-guided fuzzing framework for ARM-based IoT devices. In: Zhou, J., et al. (eds.) ACNS 2020. LNCS, vol. 12418, pp. 239–254. Springer, Cham (2020). https://doi.org/10.1007/978-3-030-61638-0_14
7. Litoussi, M., Kannouf, N., El Makkaoui, K., Ezzati, A., Fartitchou, M.: IoT security: challenges and countermeasures. Procedia Comput. Sci. **177**, 503–508 (2020)
8. Do, Q., Martini, B., Choo, K.-K.R.: The role of the adversary model in applied security research. Comput. Secur. **81**, 156–181 (2019)
9. Bansal, M., Gupta, S., Mathur, S.: Comparison of ECC and RSA algorithm with DNA encoding for IoT security. In: 2021 6th International Conference on Inventive Computation Technologies (ICICT), pp. 1340–1343 (2021)
10. Lohachab, A., Karambir: ECC based inter-device authentication and authorization scheme using MQTT for IoT networks. J. Inf. Secur. Appl. **46**, 1–12 (2019). https://doi.org/10.1016/j.jisa.2019.02.005
11. Das, A.K., Wazid, M., Yannam, A.R., Rodrigues, J.J., Park, Y.: Provably secure ECC-based device access control and key agreement protocol for IoT environment. IEEE Access **7**, 55382–55397 (2019)
12. AlMajed, H., AlMogren, A.: A secure and efficient ECC-based scheme for edge computing and internet of things. Sensors **20**, 6158 (2020)
13. Gabsi, S., Kortli, Y., Beroulle, V., Kieffer, Y., Alasiry, A., Hamdi, B.: Novel ECC-based RFID mutual authentication protocol for emerging IoT applications. IEEE Access **9**, 130895–130913 (2021)

14. Patel, C., Bashir, A.K., AlZubi, A.A., Jhaveri, R.: EBAKE-SE: a novel ECC-based authenticated key exchange between industrial IoT devices using secure element. Digit. Commun. Netw. **9**, 358–366 (2023)
15. Majumder, S., Ray, S., Sadhukhan, D., Khan, M.K., Dasgupta, M.: ECC-CoAP: elliptic curve cryptography based constraint application protocol for internet of things. Wirel. Pers. Commun. **116**, 1867–1896 (2021)
16. Hussain, S., Chaudhry, S.A., Alomari, O.A., Alsharif, M.H., Khan, M.K., Kumar, N.: Amassing the security: an ECC-based authentication scheme for internet of drones. IEEE Syst. J. **15**, 4431–4438 (2021)
17. Sowjanya, K., Dasgupta, M., Ray, S.: A lightweight key management scheme for key-escrow-free ECC-based CP-ABE for IoT healthcare systems. J. Syst. Archit. **117**, 102108 (2021)
18. Srivastava, A.,Kumar, A.: A review on authentication protocol and ECC in IOT. In: 2021 International Conference on Advance Computing and Innovative Technologies in Engineering (ICACITE), pp. 312–319 (2021).

A Hybrid Optimization Driven Deep Residual Network for Sybil Attack Detection and Avoidance in Wireless Sensor Networks

Anupama Bollampally$^{(\boxtimes)}$, Anil Kumar Bandani, and Sravani Pangolla

B.V. Raju Institute of Technology, Hyderabad, Telangana 502313, India
anupama.bollampally@bvrit.ac.in

Abstract. Small Sensor nodes that can sense, store, and send data make up a wireless sensor network (WSN). In addition, the sensor nodes are limited when it comes to energy along with computing capability. Clustering is a good way to reduce the amount of energy used by sensor nodes when transferring information to the ground station (GS) from the transmitter node. For information transmitting among nodes, multi-path routing is utilized. The WSN, on the other hand, is vulnerable to a variety of security vulnerabilities that might decrease network performance. WSNs are vulnerable to a variety of assaults, which have harmed the network's overall performance. Consequently, the fundamental objective of this study is to design and create a WSN-based hazard detection and avoidance systems. This research proposes a learning technique towards warning systems and defense in Wireless Sensor Networks that is both energy-efficient and optimization-aware. This method begins with the emulation of WSN nodes. The suggested model's overall methodology includes WSN simulation, selection of cluster heads forwarding to BS, Finally, BS has a Sybil approach to detect and assault protection. Following the WSN simulation, the LEACH Technique is used to select its most sustainable power cluster head. Such information would then be transported using the fractionated artificial bee colony (FABC) method. After the data is collected at BS, it is used to execute assault mitigation and detection activities. The Jaro-Winkler gap is employed to identify the essential feature in data. Following the identification of the ideal rules, The Sybil assault use the deep residual network (DRN). To train DRN, the Competitive Multi-Verse Rider Optimizer (CMVRO) was suggested, which blends the Competitive Multi-Verse Optimizer (CMVO) with both the Rider Optimization Algorithm (ROA). This bit rate is employed to counteract Sybil's assault. In this case, the throughput is used.

Keywords: Wireless sensor Networks · Wireless threat avoidance and detection · Sybil Attack Deep Residual Network · Avoidance of Attacks

1 Introduction

A wireless sensor network is composed of two kinds of gadgets: sensor network and sink nodes. This connection is useful for monitoring and transmitting both physical and environmental concerns to a centrally placed BS, including such pressure, temperature,

© The Author(s), under exclusive license to Springer Nature Switzerland AG 2024
S. Satheeskumaran et al. (Eds.): ICICSD 2023, CCIS 2122, pp. 145–161, 2024.
https://doi.org/10.1007/978-3-031-61298-5_12

and noise. The BS supports in information management and the display of essential actions. A WSN might be unstructured or regulated, according on the locale, with such a sensor network developed for governing regions via sensors. A regulated network is trustworthy in areas where amorphous networks are conceivable, such as streets, parking lots, buildings, and roads, as well as deserts, disaster areas, and other conditions where amorphous networks are feasible, such as the woodland. A vast area is separated into clusters in both types of networks, each of which has small sensors and CH. The cluster is made up of standard sensor nodes and sensor nodes that are joined together based on coverage capacity and it has a CH for communication with other clusters. Data from all clusters is collected and transmitted to either a mobile sink for preprocessing before being distributed to some other heterogeneous network. The CH is a critical routing strategy that enables all cluster nodes to relay the BS with data in a single or several hops. The WSN is made up of thousands of specialized nodes for sensors that are capable of processing, transmission, and storage, as well as computation. For most critical applications that WSN is expected to enable, privacy has become a major problem. The WSN's key features render them vulnerable to numerous forms of attack by definition. This paper focuses on a particularly devastating type of assault known as the Sybil attack. Sybil attacks may substantially impair network performance and network security by violating several network agreements. During this attack, a hacked node can employ many pseudo-IDs to masquerade as multiple nodes at the same time. When the Sybil assault succeeds, the attacker disables the distributed storage system, routing method, and sensor network data merging mechanism. During this attack, a hacked node can employ many pseudo-IDs to pose as multiple nodes at the same time. It does not represent any junction, but it swiftly adopts the identity of the other among numerous nodes, resulting in routing agreement redundancy. Sybil attacks undermine data integrity, security, and resource utilization.

2 Proposed ECMVRO Enabling Assault Detection Method Based on DRN

For Sybil attack prevention and detection in WSN, an efficient energy minimization conscious deep model is built. In this method, the initial step is to imitate WSN nodes. WSN modeling, cluster decision, relaying to BS, Sybil threat identification in BS, and lastly threat mitigating in BS are all part of the proposed model's overall process. When the WSN simulation is finished, a Lower Energy Adaptive Clustering Hierarchy (LEACH) methodology is used to extract the most energy efficient bunch head. The data is then routed using the FABC algorithm (fractional artificial bee colony). The accumulated data is used to perform attack detection and mitigation once the data has been collected at BS. To select the imperative characteristic from data, Jaro-Winkler distance is used. The Sybil outbreak tracking is executed with the deep residual network after all optimum rules have been determined (DRN).

The Competitive Multi-Verse Rider Optimizer (CMVRO), that integrates the Competitive Multi-Verse Optimizer (CMVO) with the Rider Optimization Algorithm, was proposed to train DRN (ROA). To counteract Sybil's attack, the data rates are used. Whenever the attacker is recognized, the communication rate is decreased. Figure 1 depicts the architecture of the designed system for attacker mitigation and detection in WSNs.

2.1 Evaluation of WSN

WSN is made up of core network and sensor nodes, each of which has its own power source, storage, transceivers, and processor. The WSN uses a variety of nodes to perceive the network. As a result, when a node is active, it can send and get information at any moment and it cannot contribute to data transmission when it is inactive. The network lifespan is seen as a significant measure for monitoring and sensing in this scenario. For a variety of causes, including network disconnection or sensor node termination, WSNs may fail to meet monitoring needs. Like a result, boosting the energy is critical for improving the lifespan of a mobile network. Acknowledge F mobile sensor nodes $f_1, f_2, \ldots, f_F$ are presented in WSN in an arbitrary manner to cover G targets $g_1, g_2, \ldots g_G$. Each sensor node has an initial energy and the ability to manage itself for data transmission. The range of sensing is $l_1, l_2, \ldots l_q$ based on energy consumption $m_1, m_2, \ldots, m_q$. Furthermore, the BS is within network area of every sensor. The purpose this time is to assemble a group of sensors in such a way that it can detect targets.

2.2 Cluster Head Selection Using LEACH Protocol

Clustering is a clever solution to the energy dilemma presented by rechargeable batteries sensor nodes in WSNs. In order to address the energy concerns, the LEACH algorithm is used to pick CH. LEACH is utilized to cluster nodes so that CH is picked using efficient sensor nodes. The LEACH approach is used to pick cluster heads, which link sensor nodes collectively based on their commonalities rather than having them difficult into a single cluster. The LEACH approach [1, 2] creates an intense sensor network with equivalent networks to transmit data to BS. As a result, the optimal CH for data gathering and delivery to the BS is chosen. CH consumes more energy to enable contact between the nodes because of sink network is situated in a remote area. As an outcome, the LEACH technique facilitates in the choice of the most suitable CH, resulting in a higher energy content for the chosen node. As just a corollary, the LEACH protocol employs randomized CH rotations to distribute energy evenly among sensor nodes. Furthermore, the LEACH is structured in a specific way, and once CH is found, then LEACH operates in $\frac{1}{t}$ rounds. A pair of CH is identified using size for each round. wt, where w symbolize complete nodes in t rounds. Advertisement, cluster configuration, and broadcast scheduling are the three sub-phases of the setup process. During the advertisement phase, every station produces an arbitrary value and computes it using a specific edge value. Also, criterion is written in the following format:

$$\tau(w) = \begin{cases} \dfrac{t}{1 - t \times (c \bmod \frac{1}{t})} & ; if \ e \in \alpha \\ 0 \ ; Otherwise \end{cases} \tag{1}$$

where, α symbolize node, which is not CH, t express CH percentage, and c signifies update period of recent topology. As either a consequence, stations which have chosen a

CH send advertisement packets to their nearby nodes. Whenever a cluster is formed, all sites in the system react to a CH announcement informing them of their choices. During the broadcast phase, all node responses are compiled in order to make a judgement on cluster membership. As a result, the CH constructs a TDMA agenda based on the batch's total number of stations. This agenda smacks a node-to-node agreement on the timing for broadcasting messages at a specific time. Finally, data is accumulated in CH for data transfer, and the generated data is supplied to the base station. As a result, LEACH's CH is stated as,

$$K = \{K_1, K_2, \cdots, K_g, \cdots, K_o\} \; ; \; 1 \leq g \leq o \text{ where, } o \text{ Symbolize total CH} \quad (2)$$

2.3 Routing Using FABC

The routing is done among many of the cluster heads created during the clustering phase. Routing is crucial in a WSN because it allows the transmission of data between the network system and the BS. Data transport between stations and the BS requires routing in order to establish communication. To begin relaying and transfer the gathered data to the BS, the FABC [3] approach is employed. The FC and ABC algorithms are combined to generate the FABC algorithm. Conversely, ABC [4] is an ultimately improve that is motivated by cognitive foraging behavior. FC [5] is a well-known mathematics expansion for analyzing the best answers of earlier installments and upgrading the present iteration's solutions. As both a consequence, FC and ABC have joined forces to broaden the search for solutions. Like a consequence, the FABC contributes toward the settlement of exploration and utilization problems, as well as increased global data use. The FABC model's new routing technique phases are as follows:

Trace 1: Loading
Take up Q_R as foodstuff origin loaded in a random manner along the foodstuff origin is considered to be $Q_R \times g$. The parameters of integer are repeated in the patterns in an arbitrary fashion feeding at intervals 1 to p. The power is assigned to every node and the sensor stations are installed in the network. The sink node has access to each node's location information in order to perform path linking.

Trace 2: Employed Bee
Every foodstuff origin is ciphered in FABC, as well as every foodstuff origin characteristic is shown in the process of a sensor station basis, with distance of the foodstuff origin considered into balanced every number about CH in IOT. The bee's food source is changed, and fresh bee stage development is predicated upon ABC [4] and used the equation,

$$Q_{u,v}^{s+1} = Q_{u,v}^s + A_{u,v}(Q_{u,v}^s - Q_l, v) \quad (3)$$

where, $Q_{u,v}^s$ represents u^{th} food source of vth value in s^{th} iteration, $A_{u,v}$ indicates the arbitrary parameter for the range of [-1,1], $u \in \{1, 2, \ldots, g\}$ where l indicates the indicator of neighbor in alike a way a certain $l \in \{1, 2, \ldots, Q_R\}$.

The resultant equation is stated as, later reconstructing the actual foodstuff origin to enhance the result offshoot structure.

$$Q_{u,v}^{s+1} - Q_{u,v}^{s} = A_{u,v}(Q_{u,v}^{s} - Q_l, v) \tag{4}$$

Here, $Q_{u,v}^{s+!}$ represents the distinct type of offshoot with structure $\alpha_Q = 1$ is defined by,

$$D^{\alpha_Q}\left[Q_{u,v}^{s+1}\right] = A_{u,v}(Q_{u,v}^{s} - Q_l, v) \tag{5}$$

If the FC [5] perception is used, the structure of food origin is designed in an original figure, resulting in simple modification as well as prolonged recollection chattels. As a result, each of the above statements can be expressed as, where two sites of differential outgrowth are assumed.

$$Q_{u,v}^{s+1} - \alpha_Q Q_{u,v}^{s} - \frac{1}{2}\alpha_Q Q_{u,v}^{s-1} = A_{u,v}(Q_{u,v}^{s} - Q_l, v) \tag{6}$$

$$Q_{u,v}^{s+1} = \alpha_Q Q_{u,v}^{s} + \frac{1}{2}\alpha_Q Q_{u,v}^{s-1} + A_{u,v}(Q_{u,v}^{s} - Q_l, v) \tag{7}$$

$$Q_{u,v}^{s+1} = |\alpha_Q Q_{u,v}^{s} + \frac{1}{2}\alpha_Q Q_{u,v}^{s-1} + A_{u,v}(Q_{u,v}^{s} - Q_l, v)| \tag{8}$$

The development structures the driven restraints being extracting the value of the result in a certain integer interval after the production of a new food source.

Trace 3: Evaluation of Fitness Function
The acquired foodstuff origin is assumed along the strength function. Assuming strength about enhanced foodstuff origin $Q_{u,v}^{s+1}$ is limited when compared with elderly foodstuff origin $Q_{u,v}^{s}$. Now, the response is enhanced by elderly finest response $Q_{u,v}^{s}$ either if not, the answer is established about every newer foodstuff origin $Q_{u,v}^{s+1}$.

Trace 4: Spectator Bee
With every observer bee aspect, every foodstuff origin of the other part away from every community are enhanced, and every foodstuff origin act chosen using the statement,

$$y_u = \omega_1 \times \frac{fit_u}{Max_{u=1}^{Q_R} fit_u} + \omega_2 \tag{9}$$

where, ω_1 and ω_2 represent stable along with fit_u shows strength activity hired as selecting the finest route. Every selected foodstuff origin is restored along current answer $Q_{u,v}^{s+1}$. Every bee inspects as long as every current foodstuff origin if strength of $Q_{u,v}^{s+1} < Q_{u,v}^{s}$, otherwise remains same. The fitness of the FABC algorithm is influenced by three factors: distance, energy, and delay. When selecting a path, keep in mind that the length is deemed to be short, the power of the mobile stations along the route must be greater, and the latency should be as low as possible. As a result, these design constraints are employed to choose a path. As a result, the goal is to reduce the objective function, which is represented by,

$$fit_u = \eta_1 h_u^{loc} + \eta_2 h_u^{energy} + \eta_3 h_u^{delay} \qquad (10)$$

where, η_1, η_2 and η_3 denote weighted(entire) parameters, and h_u^{loc} represent every length of assemblage associate to CH. Furthermore, h_u^{energy} indicates the power of the stations involved in a route along with h_u^{delay} is the time lag incurred in every transmission.

Trace 5: Detective Bee
If there have been no changes in foodstuff supplies in past cycles, these phases will be adopted. The selected foodstuff origin act rejected and recovered through a freshly obtained foodstuff origin at arbitrarily.

Trace 6: Completion
The preceding stages are continued until the maximum number of cycles is reached, Z_{max}. Explanation response continue the finest foodstuff origin. Following every selection of the shortest pathways, communication among CH along with the BS is logged away for the purpose of exchanging sensor node information.

2.4 Attack Detection Using Proposed ECMVRO-DRN

The assault detection has been completed at BS. DRN [6] is used in this circumstance for detect that an intruder or a genuine user is present. The DRN is composed of layers that include activation functions, convolution groups, a continuously segregator, and equilibrium merging bands. The design of a deep residual network is seen in Fig. 4.

Convolution (Conv) Layer
The input data is processed by the convolution layer using a set of filters known as kernels that take into account local connections. The convolution layer's computation procedure is as follows:

$$B2d(M) = \sum_{a=0}^{E-1}\sum_{s=0}^{E-1} X_{a,s} \bullet M_{(u+a),(v+s)} \qquad (11)$$

$$B1d(M) = \sum_{Z=0}^{C_{in}-1} G_Z * M \qquad (12)$$

where, M denote CNN feature of load figure, u along with v act utilized as long as recoding correspondents, G symbolize $E \times E$ kernel model along with is likewise known as gain value, a along with s denote location catalogue of kernel model. Thus, G_Z symbolize extent of kernel as Z^{th} aid neuron, along with $*$ denote vexed alternation driver.

Merging Layer: On every cut as well as extent of the factor graph, average merging is chosen to function.

$$a_{out} = \frac{a_{in} - Z_a}{\lambda} + 1 \tag{13}$$

$$s_{out} = \frac{s_{in} - Z_s}{\lambda} + 1 \tag{14}$$

where, a_{in} signifies input matrix width, s_{in} symbolize peak of input model, a_{out} and s_{out} are the corresponding response(output) values. Furthermore, Z_a as well as Z_s signifies width and height of kernel size. Non-continuous process simulation can be used to learn non-continuous and difficult aspects, as well as to improve the non-continuity of mined data. The ReLU function is written like this:

$$\text{Re}LU(M) = \begin{cases} 0 \; ; \; K < 0 \\ K \; ; \; K \geq 0 \end{cases} \tag{15}$$

Here, K signifies feature.

Batch convergence: To boost training reliability and speed, the input bands are broadened by changing and grading the simulations.

Residual blocks: To meet input with output for different sizes, the element identical point is employed.

$$O = \Re(M) + M \tag{16}$$

$$O = \Re(M) + \lambda_M M \tag{17}$$

where M and O quote input and output enduring sections, O signifies aligning rapport, the λ_H denote element identical point.

Straight divider runs the process of determining strident pixels against the input picture once the convolution floor is finished. It combines a fully connected layer with a soft max function.

$$O = \lambda O + \upsilon \tag{18}$$

Here, λ denote weight matrix, and υ indicate bias. DRN's architectural model is shown in Fig. 2. The DRN result is displayed as O in this case, which aids in determining in case the user is an attack or a typical user.

2.5 Guiding of (DRN) Deep Residual Network

The ECMVRO Technique, which has been suggested, will be used to guide deep residual networks. The proposed ECMVRO will be used to direct the weight of the analyzer (classifier) in order to provide the best solution. ECMVRO enhances the deep residual network by bringing together ROA [7], CMVO [8] to select the best loads (weights) for active steering of the classifier's in-house design values. The steps of the ECMVRO Technique are outlined below.

Method 1. Loading
The early phase is response loading, which involves setting up the solutions as well as other parameters like iteration count. The solutions are written as follows:

$$G = \{G_1, G_2, \ldots, G_s, \ldots, G_t\} \tag{19}$$

where, t signifies total solutions, G_s signifies s^{th} solution.

Method 2. Computation of Failure
The best answer is found by considering a failure role, and the failure role is considered as a denigration issue, thus the answer with the lowest MSE is chosen. Finally, as result, MSE is evaluated in the following Method:

$$MS_{err} = \frac{1}{g} \sum_{h=1}^{g} [\xi_h - O]^2 \tag{20}$$

where, ξ_h symbolize expected output and O express output generated from the combination of DRN and NN classifier, g refers count of data samples, such that $1 < h \le g$.

Method 3. Discovering Update Position of Riders
To find the leader, the position of each rider in each set is updated. As a result, the procedure for updating a rider's position using the unique feature of each rider is given below. Each rider's current position is listed below.

(a) Position update based on bypass rider
  As per ROA [7], the bypass riders pose a recognizable path and its update position is expressed as,

$$L_{n+1}^B(w, u) = \alpha[L_n(\beta, u) * \eta(u) + L_n(\rho, u) * [1 - \eta(u)]] \tag{21}$$

where, α symbolize random number, β signifies arbitrary number amongst 1 to P, ρ denote a arbitrary number in 1 to P and η express arbitrary number between 0 and 1.Assume $\beta = w$,

$$L_{n+1}^B(w, u) = \alpha[L_n(w, u) * \eta(u) + L_n(\rho, u) * [1 - \eta(u)]] \tag{22}$$

$$L_{n+1}^B(w, u) = \alpha L_n(w, u) * \eta(u) + \alpha L_n(\rho, u) * [1 - \eta(u)] \tag{23}$$

  The CMVO is effective at improving computational efficiency and convergence rate speed. The update equation is provided as,according to CMVO [8].

$$L_{n+1}^B(w, u) = o_1 * TDR + o_2 * (L_n(s, u) - L_n(w, u)) + o_3 * (L_u - L_n(w, u)) \tag{24}$$

  The successful networking procedure is carried out over here employing a traffic-aware routing scheme built on Fractional Glow-Worm Swarm Optimization (FGWSO).

Sybil attack detection is carried out at the base station [9]. Where, $L_n(s, u)$ refers winner universe for u^{th} iteration of race, $L_n(w, u)$ signifies loser universe for u^{th} iteration of race, L_u Symbolize mean position value of relevant universe, TDR is coefficient and $o_1, o_2\ o_3$ signifies random numbers between [0, 1]. The TDR is given as,

$$TDR = 1 - \left(\frac{n^{1/\ell}}{N^{1/\ell}} \right) \tag{25}$$

where, $\ell = 6$, current iteration is represented as n, then maximum iteration is expressed as N.

$$L_{n+1}^B(w, u) = o_1 * TDR + o_2 * L_n(s, u) - o_2 * L_n(w, u) + o_3 * L_u - o_3 * L_n(w, u) \tag{26}$$

$$L_{n+1}^B(w, u) = o_1 * TDR + o_2 * L_n(s, u) + o_3 * L_u - L_n(w, u)(o_2 + o_3) \tag{27}$$

$$L_n(w, u)(o_2 + o_3) = o_1 * TDR + o_2 * L_n(s, u) + o_3 * L_u - L_{n+1}^B(w, u) \tag{28}$$

$$L_n(w, u) = \frac{o_1 * TDR + o_2 * L_n(s, u) + o_3 * L_u - L_{n+1}^B(w, u)}{o_2 + o_3} \tag{29}$$

Substitute Eq. (24) in Eq. (19),

$$L_{n+1}^B(w, u) = \alpha \left(\frac{o_1 * TDR + o_2 * L_n(s, u) + o_3 * L_u - L_{n+1}^B(w, u)}{o_2 + o_3} \right) * \eta(u)$$
$$+ \alpha L_n(\rho, u) * [1 - \eta(u)] \tag{30}$$

$$L_{n+1}^B(w, u) = \frac{\alpha o_1 * TDR + o_2 * L_n(s, u) + o_3 * L_u}{o_2 + o_3} * \eta(u)$$
$$- \frac{L_{n+1}^B(w, u)}{o_2 + o_3} \alpha * \eta(u) + \alpha L_n(\rho, u) * [1 - \eta(u)] \tag{31}$$

$$L_{n+1}^B(w, u) + \frac{L_{n+1}^B(w, u)}{o_2 + o_3} \alpha * \eta(u) = \frac{\alpha o_1 * TDR + o_2 * L_n(s, u) + o_3 * L_u}{o_2 + o_3} * \eta(u)$$
$$+ \alpha L_n(\rho, u) * [1 - \eta(u)] \tag{32}$$

suggested safety geolocation detection and detection against numerous threats based on an optimal multilayer perceptron artificial neural network (MLPANN). The suggested approach is divided into two sections: localization techniques and machine learning techniques for detecting and localizing WSN DoS attacks [10].

$$L_{n+1}^B(w, u) \left(1 + \frac{\alpha * \eta(u)}{o_2 + o_3} \right) = \frac{\alpha o_1 * TDR + o_2 * L_n(s, u) + o_3 * L_u}{o_2 + o_3} * \eta(u)$$
$$+ \alpha L_n(\rho, u) * [1 - \eta(u)] \tag{33}$$

$$L_{n+1}^{B}(w, u)\left(\frac{o_2 + o_3 + \alpha * \eta(u)}{o_2 + o_3}\right) = \frac{\alpha o_1 * TDR + o_2 * L_n(s, u) + o_3 * L_u}{o_2 + o_3} * \eta(u)$$
$$+ \alpha L_n(\rho, u) * [1 - \eta(u)] \tag{34}$$

Last concluding statement of projected CMVRO is stated in the following manner,

$$L_{n+1}^{B}(w, u) = \frac{o_2 + o_3}{o_2 + o_3 + \alpha * \eta(u)}$$
$$\left[\frac{o_1 * TDR + o_2 * L_n(s, u) + o_3 * L_u}{o_2 + o_3}\alpha * \eta(u) + \alpha L_n(\rho, u) * [1 - \eta(u)]\right] \tag{35}$$

By detecting Sybil and buffer overflow attacks across IoT device architecture, a deep learning technique is utilized to discover illegal Connected technologies. The CNN approach focuses on identifying any malevolent node's threat or behavior and tries to address the issue [11].

Method 4: Riding Off Time
The procedures are carried out again till the timer runs out, at which point the attacker is assessed. Table 1 shows pseudo code for the developed CMVRO. A mechanism is presented for identifying Sybil assaults in the network. Sybil attacks have been discovered by comparing the properties of each node to those of network nodes [12]. an overview of the most up-to-date assured strategies to protect out from Sybil assault. Mitigation strategies for the Sybil assault involve cryptography, trust, received signal indicator (RSSI), cryptography, and artificial intelligence [13].

A single method has previously been put out for a secure platoon management system that can fend off Sybil attacks and safeguard significant events that take place during platoon operations. By implementing key exchange, digital signature, and encryption algorithms utilizing elliptic curve cryptography (ECC), the vehicle identification and message sent are both authenticated [14]. The goals of the cluster-based hierarchy routing protocol LEACH-E (Low Energy Adaptive Clustering Hierarchy-Energy) are to maintain the original protocol's functionality and offer safe routing. A cluster head (CH) is always chosen by this energy-efficient approach based on the cluster group's high energy level. In this instance, we suggest using a LEACH-E-GA to identify intrusions (ID) in the nodes of wireless sensors. To avoid Sybil assaults, the genetic algorithm is implemented in LEACH-E [15]. For the purpose of detecting Sybil attacks in wireless sensor networks (WSNs), the system implemented a policy based on the strength of the received signal indicator (RSSI). A suitable threshold was selected to initiate the Sybil attack detection approach in the event that an attack is very likely to occur. The simulation's findings demonstrate that the system can efficiently identify Sybil attacks while using little energy [16].

[17] suggested a detection method for the Sybil Attack; in contrast to the other method, this method is time- and situation-efficient. Use the LEACH routing algorithm in WSN to identify and stop Sybil attacks. Length and hop count between nodes are utilized for detection, and the Breadth First Search (BFS) Protocol is employed as an encryption mechanism to thwart Sybil attacks [18].

Table 1. Methods of Sybil Attack

Sl. No.	Method of Sybil Attack	Description
1	Using Fake Identity to Perform a Sybil attack	In the fake identity method, the attacker uses many pseudonymous identities to get influenced by the VANET on a massive scale
2	Using Identity Theft to perform a Sybil attack	In this attack, the Attackers use the stolen information from the accessible sources, access the vehicular network, join the network, and do wicked things
3	Conspired Sybil Attack, sock puppets	An identity used for description is a sock puppet. It portrays conversations that are used for the malicious attack but looking useful. In this method, the attackers of the Sybil attack victim the stakeholders to gain access to the vehicular ad hoc network
4	Compromise on Message Integrity	Message/data consistency ensures that it has not been modified in any manner, like insertion or delete replay attacks or by frame addition or deletion. In this assault, the messages transmitted from one node to another network node are comprised and modified to trigger the attack
5	Insider Attacker	In an insider attack, an entity resides in the network with a harmful aim. Due to the reason that it uses the inside information about the system, its detection is not an easy task, and, in this way, the attacker negatively impacts the services of the vehicular network

2.6 Attack Mitigation with Data Rates

If the last response O is attacker, The attack is then mitigated. To safeguard the connection, application, and server, IT managers deploy a monitoring and preservation strategy that limits the effect of malicious activities and threats while retaining user functionality.

Data rate is reduced when the attacker is identified. For allotted time period Δt, let $L(\Delta t)$ represent total packets, l_k represent data packet size of kth data packet and l_{max} represent maximal packet size. The Average mean of packet size is represented as,

$$\vec{l} = \frac{1}{c}\sum_{k=1}^{c} l_k; \; 60 \leq \vec{l} \leq l_{max} \tag{36}$$

If $\vec{l}$ of a user is beyond the limit, then minimize the data rate to 50%.

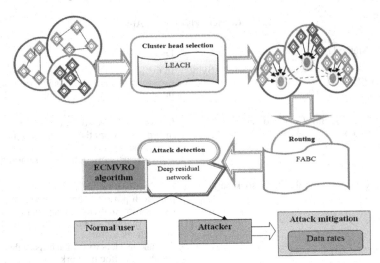

Fig. 1. Block Diagram of Suggested Model for Detection and Avoidance of Sybil Attacks in WSN

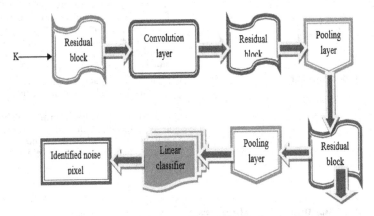

Fig. 2. Block Diagram of Deep Residual network

2.7 Developed Sybil attack detection Model-Flow Chart

The LEACH system is a routing system that separates the sensor field into numerous clusters. Each cluster contains one node that serves as the cluster head. After aggregation, the cluster head gets data from the cluster's member nodes and delivers it to the base station. The LEACH Protocol is divided into two phases: a set-up phase that determines the cluster structure and distribution schedule, and a steady-state phase that transmits the information defined during the set-up phase. The cluster head consumes a lot of energy since it receives data from the cluster's member nodes and then sends it back to a base station after aggregation. An artificial bee colony's ABC algorithm. The built-in system's goal is to assist in the design of the best route between the specified beginning location and a single of the numerous destination places. The procedure was updated in this experiment by introducing an alternative function on one of the equation's components. For this purpose, artificial bees were classified as scout bees, active bees, and observing bees. Then, for the observing bees, an additional function was introduced. Apart from looking for novel solutions, these bees seek to shorten already-remembered paths. To train DRN (ROA), the Competitive Multi-Verse Rider Optimizer (CMVRO) was suggested, which blends the Competitive Multi-Verse Optimizer (CMVO) with the Rider Optimization Algorithm. The information's rates are employed to counter Sybil's assault. When the attacker is identified, their interaction rate is reduced (Fig. 3).

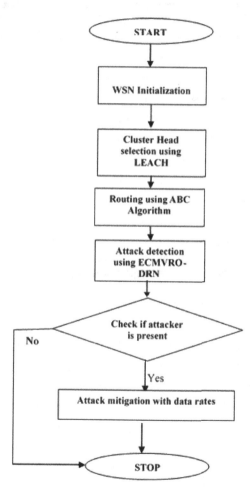

Fig. 3. Proposed Sybil attack detection and mitigation model in WSN flow chart

3 Simulation Results

The Proposed work is compared with ABC Algorithm (Artificial Bee Colony), Leach Protocol. The Simulation is Performed by using NS-2 Tool and Gnu plotting and the parameters delay, energy Consumption, packet loss, end to end delay and throughput is evaluated. The Proposed Protocol ECMVRO is evaluated and found to be efficient (Figs. 5, 6, 7, and 8).

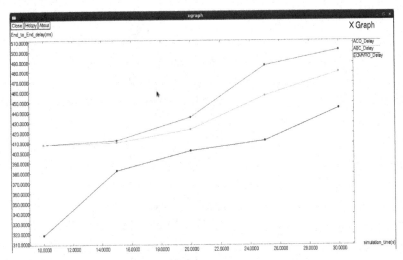

Fig. 4. Techniques with a delay are evaluated.

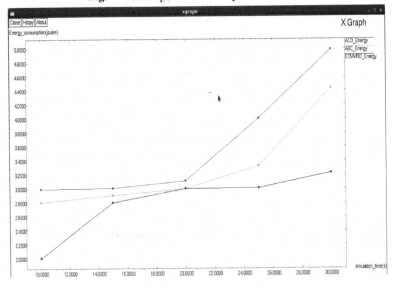

Fig. 5. Energy Consumption Vs. Simulation Time

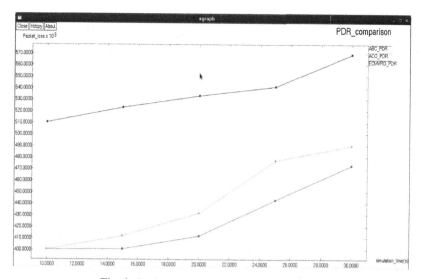

Fig. 6. Packet Loss (Lag) Vs. Simulation Time

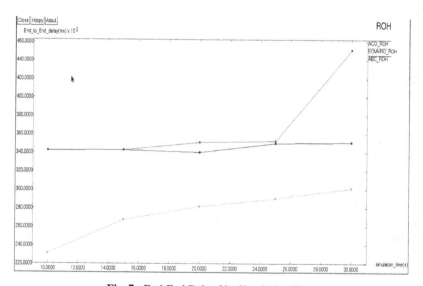

Fig. 7. End-End Delay Vs. Simulation Time

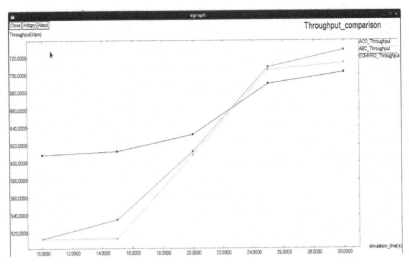

Fig. 8. Throughput Vs. Simulation Time

4 Conclusion

For Sybil attack prevention and detection in WSN, an efficient energy optimize conscious deep model is built. In this method, the initial step is to imitate WSN nodes. WSN modeling, cluster member's choice, forwarding to BS, Sybil threat identification in BS, and ultimately threat mitigation in BS is all part of the proposed model's overall process. Following the completion of the WSN model, the LEACH approach is used to determine its most efficient energy cluster head. The data is then transported using the FABC method. Once the data has been acquired at BS, it is used to execute attack detection and mitigation. The Jaro-Winkler distance is employed to identify the essential feature in data. The Sybil attack monitoring is figured out using the deep recurrent neural network after the optimal parameters have been discovered (DRN). The Competing Multi-Verse Rider Optimizer (CMVRO), that incorporates the Competitive Multi-Verse Optimizer (CMVO) with the Rider Optimization Algorithm, was proposed to train DRN (ROA). To counteract Sybil's attack, the data rates are used. Whenever the attacker is detected, the communication rate is decreased.

References

1. Ramesh, S., Rajalakshmi, R., et al.: Optimization LEACH protocol in wireless sensor network using machine learning. Comput. Intell. Neuro Sci. **2022**, 1–8 (2022)
2. Panda, J., Indu, S.: Localization and detection of multiple attacks in wireless sensor networks using artificial neural networks. Wirel. Commun. Mobile Comput. (2023). https://doi.org/10.1155/2023/2744706
3. Cui, Y., Hu, W., Rahmani, A.: FABC-Fractional order artificial bee colony algorithm with application in robot path planning. Europ. J. Oper. Res. **306**(1), 47–64 (2023)
4. Famila, S., Jawahar, A.: Improved artificial bee colony optimization-based clustering technique for wireless sensor networks. Wirel. Pers. Commun. **110**(9), 1–10 (2020)

5. Sawwa J.A., Almseidin, Md., A.: Spark based artificial Bee Colony algorithm for unbalanced large data classifications. MDPI Information **13**(11), 530 (2022)
6. Zhou, Q., et al.: Training deep learning neural network for wireless sensor networks, using loosly and weakly labelled images. Neuro Comput. 1–11 (2020)
7. Susan, A., Ananth, J.P.: A modified rider optimization algorithm for multihop routing in WSN. Int. J. Numer. Model. Electron. Netw. Fields **33**(1) (2020)
8. Ilyas, B., Xie, K., Mouna, C.: A new competitive multiverse optimization technique for solving objective and multi objective problems. Eng. Rep. **2**(1), 1–33 (2020)
9. Neetha, C.V., Anitha, A., Mukesh, M.: An optimisation driven deep residual network for Sybil attack detection with reputation and trust-based misbehaviour detection in VANET. J. Exp. Theor. Artif. Intell. (2022)
10. Gebremariam, G.G., Panda, J., Indu, S.: Localization, and detection of Multiple attacks in wireless sensor networks using artificial neural networks. Wirel. Commun. Mobile Comput. Hindawi **2023**(2744706), 1–29 (2023)
11. Pandey, U., Kumar, B.C.: IoT edge based frame work for Sybil and buffer overflow detection. Res. Square 1–17 (2022)
12. Younis, M.I., Latif, R.M.A., Haq, I.: An evaluation of Sybil attacks detection approaches in VANETs. **10**(25) (2022)
13. Arshad, A., Hanapi, Md.Z., Subramaniam, S., Latip, R.: A survey of Sybil attack counter measures in IoT based wireless sensor networks. Peer J. Comput. Sci. (2021)
14. Junaidi, D.R., Ma, M., Su, R.: Secular vehicular platform management against Sybil attacks. Sensors (Basel) **22**(22), 9000 (2022)
15. Amruthavalli, R., Bhuvaneshwaran, R.S.: Genetic Algorithm Enabled Prevention of Sybil Attacks for LEACH-E. Mod. Appl. Sci. **9** (2015). https://doi.org/10.5539/mas.v9n9p41
16. Chen, S.-S., Yang, G.: LEACH protocol based security mechanism for Sybil attack detection, ongxin Xuebao/J. Commun. **32**, 143–149 (2011)
17. Kumari, D., Singh, K., Manjul, M.: Performance evaluation of Sybil attack in cyber physical system. Procedia Comput. Sci. **167**, 1013–1027 (2020). https://doi.org/10.1016/j.procs.2020.03.401
18. Deepak, P., Archana, Rani, P.: Implementation of optimized LEACH routing protocol for detection and prevention of Sybil attack in wireless sensor networks using BFS. Int. J. Innov. Sci. Eng. Technol. **3**(6) (2016)

Feature Engineering Techniques for Stegware Analysis: An Extensive Survey

M. Anitha[1]([✉])([iD]) and M. Azhagiri[2]([iD])

[1] Department of Computer Science and Engineering, SRM Institute of Science and Technology, Ramapuram, Chennai, India
anithamuthulingam.26@gmail.com

[2] Department of Computer Science and Engineering, SRM Institute of Science and Technology, Ramapuram, Chennai, India
azhagirm@srmist.edu.in

Abstract. The art of hiding secret text within an innocuous cover medium is steganography. Steganalysis is the counterpart of steganography which focuses on the detection and extraction of the secret text from the medium. Feature engineering is the crucial field in Stegware Analysis which intends to identify more specific features, focusing on the accuracy and efficiency. Feature Engineering is a process in Machine learning where the features of any dataset are selected and extracted for further use. Feature engineering is the process of extracting, transforming and selecting the most relevant features form the data that aids in discriminating between the stego and cover image. This is because, most of the time, the data will be in a raw format. Any ML model needs the data to be pre-processed and kept ready to train the model. Thus, from the pool of raw data, the required data needs to be selected and can be used in training the model. Further, the data at point needs to be extracted to get the precise data. The scope of the work is to identify the various feature engineering techniques available in practice and efficiently use them to achieve high accuracy and precision in the system. The survey focuses on the several feature selection and extraction techniques like filter method, wrapper method and embedded methods. Correlation being one of the feature selection methods is focused; while statistical moments computes the mean, variance and skewness of the feature. The extraction method holds the Computation of Invariants and other such. Comparative study is made on both the methods to understand the concepts with ease. The work starts by taking a sample from the dataset and few feature extraction techniques are applied on the same. Then the original image is compared with the extracted images with the view of histogram. The paper gives valuable insights into the effectiveness of different feature engineering techniques using the dataset and underscores the importance of feature engineering in enhancing machine learning model performance. The selection and combination of these features can remarkable impact the effectiveness of the data, improving the accuracy of the systems.

Keywords: Correlation · Skewedness · Wrapper method · Correlation

© The Author(s), under exclusive license to Springer Nature Switzerland AG 2024
S. Satheeskumaran et al. (Eds.): ICICSD 2023, CCIS 2122, pp. 162–174, 2024.
https://doi.org/10.1007/978-3-031-61298-5_13

1   Introduction

Steganography, the art of concealing information within digital media, has gained significant importance in today's digital world. With the increasing prevalence of communication and data sharing through digital channels, the need for covert information exchange has also risen. Stegware, which refers to software or techniques used for steganographic purposes, plays a vital role in facilitating the hiding and retrieval of hidden data within various types of digital media, such as images, audio files, or videos. The significance of stegware lies in its ability to embed secret information within seemingly innocuous media, making it challenging to detect by unauthorized parties. This covert communication can be utilized for various purposes, including covert messaging, information leakage, or digital watermarking for copyright protection. However, the existence of stegware poses a significant challenge for digital forensics and security experts, as it requires sophisticated techniques to uncover the hidden data. Steganalysis, which focuses on detecting and analyzing steganographic content, feature engineering, plays a crucial role. Feature engineering is the transformation of raw data into meaningful and informative features that helps in training the machine learning models or performing statistical analysis. In steganalysis, feature engineering techniques aim to extract or select features that are sensitive to the presence of hidden data or modifications made by stegware. This paper aims to provide an overview of the feature engineering process in stegware. It will delve into the various techniques employed for feature selection and extraction and their significance in steganalysis. By understanding the goals and methods of feature engineering in stegware, researchers and practitioners in the field of steganalysis can gain insights into effective strategies for detecting and analyzing hidden information within digital media. This paper will include sections dedicated to feature selection techniques, feature extraction methods, a comparative analysis of different approaches, and challenges in feature engineering for stegware. It will also discuss future directions and emerging trends in the field.

Steganalysis or Stegware Analysis is the process of detecting the presence of hidden information within a digital image, audio file, or video file. The goal of steganalysis is to identify the existence of a covert message or payload that has been embedded within a cover medium (e.g., an image or audio file) using steganography [9]. There are different techniques and methods used in steganalysis, and they can be classified into two main categories: statistical and heuristic. Statistical methods are based on statistical analysis of the stego object (the cover object that contains the hidden message) and the corresponding original object (the unmodified object before embedding). Statistical techniques involve examining features such as the distribution of pixel values, correlation between pixels, and other statistical properties to identify the presence of a hidden message. Heuristic methods, on the other hand, are based on heuristics or rules of thumb that are derived from the analysis of known steganography techniques. These methods involve analysing specific characteristics of the stego object, such as the presence of certain file headers or metadata, to detect the presence of steganography. Steganalysis can be used to detect steganography in a variety of contexts, such as law enforcement investigations, intelligence gathering, and computer forensics. It can also be used by individuals and organizations to protect against steganography-based attacks, such as stegware, which can be used to steal information or take control of a system. As steganography techniques become more advanced, steganalysis must also continue to evolve to detect and prevent the use of steganography for malicious purposes. This requires on-going research and development of new steganalysis techniques and tools, as well as the continued use of traditional methods.

2 Related Study

The author Monika A and Eswari R [1] presents a research study focused on the detection of information hiding malware attacks using stegomalware detection system. The authors explore the growing concern of information hiding malware attacks, where malicious software hides sensitive information within innocuous files. Such attacks pose significant risks to data security. In response, the authors propose an ensemble-based detection system that combines multiple classifiers to improve the accuracy and efficacy of the process. The proposed detection system leverages various machine learning algorithms and features to identify hidden malware. The authors experiment with different classifiers, that includes Random Forest, SVM, KNN, and estimate the performance metrics such as accuracy, precision, recall, and F1-score. The ensemble approach combines the outputs of these classifiers to make a final decision on whether the analyzed file contains hidden malware. The result demonstrates the ensemble-based approach achieves greater accuracy and detection rates compared to individual classifiers. The system exhibits promising results in detecting information hiding malware attacks.

The author Boeschoten et.al, [2] proposes a novel solution for detecting stealthy content embedded in JPEG files. The approach leverages a deep learning-based model, specifically the VGG-16 architecture, to extract relevant features from JPEG images. A huge dataset containing benign and malicious JPEG images is collected and labeled for training the model. Various image processing techniques are applied to create diverse versions of the images, enhancing the system's robustness. Experimental results demonstrate that MalJPEG achieves high detection rates for malicious JPEG images while possessing a low false-positive rate, outperforming existing methods. The paper gives the potential of machine learning and deep learning in addressing the security challenges associated with image-based malware. It also discusses the limitations of the MalJPEG system and suggests future research directions for further improvement.

A case study on malware classification is focused by Gibert et.al, [3] where the authors propose a hybrid approach that combines traditional feature engineering techniques with deep learning methods. They aim to refine the accuracy and efficiency of malware classification. The study utilizes a dataset of malware samples and employs traditional feature engineering techniques, such as n-grams, entropy, and byte histograms, to extract relevant characteristics from the samples. These stay as inputs to train machine learning classifiers. The authors incorporate Convolutional Neural Network (CNN) architecture to automatically learn feature representations from raw binary data. The deep learning model is trained on the malware dataset to classify samples into different categories. The performance of their hybrid approach with individual feature engineering and deep learning methods is compared. Evaluation metrics, including accuracy, precision, recall, and F1-score, are used to assess the classification results.

The rapid tumor detection method is proposed by Yadav et.al, [4] for the unmanageable growth of the brain cells. ML methods and feature engineering techniques results in the accurate detection of the cancer cells. An ensemble based feature vectors (FV) on the GLGM and VGG16 is proposed. Classification algorithms like SVM, KNN are used in the proposed framework. Accuracy is on the higher standard of 99% on the ensemble method. Two datasets are used in this work. Since the reliability and efficacy are on the higher standard, radiologists tend to use it in the detection of the brain tumors through MRI. These experimental results can be deployed in real time.

The paper "Deep learning-based spatial domain steganalysis using convolutional neural networks" by Zhu, Huang, and Liu presents a specific approach for steganalysis in the spatial domain using Convolutional Neural Networks (CNNs). A CNN model on a dataset of images is proposed with hidden and non-hidden information to learn relevant features and classify them accurately. Through experimental evaluation, the proposed method demonstrates superior performance compared to traditional steganalysis techniques, achieving high accuracy in detecting hidden information. The paper contributes to the field by showcasing the effectiveness of CNNs for spatial domain steganalysis, providing valuable insights into the application of deep learning in steganography detection.

3 Overview of Feature Engineering

Feature engineering is transforming the raw input data into a meaningful group of data available for training a ML model is the crucial step. This way, the important features are captured and used in the stegware analysis process. This is a vital technique by providing input that effectively discriminates between cover media and stego media. The main objectives of feature engineering in stegware include feature selection and feature extraction. Table 1 describes the best Feature Engineering Techniques. The various approaches are listed along with the specific technique. Further descriptions are given.

Feature selection aims to identify a subset of relevant features from primary features that are sensitive to the presence of hidden data or modifications introduced by stegware. The selected features [2] should possess discriminatory power and enable accurate differentiation between cover and stego media. Thus, by minimizing the dimensionality of the feature space, feature selection helps to enhance the performance of steganalysis algorithms by eliminating irrelevant or redundant features.

Feature extraction, on the other hand, focuses on transforming the original data into a new representation that captures essential information for steganalysis. This process involves extracting discriminative features from the media, such as statistical measures, transform coefficients, or spatial characteristics, which can reveal patterns or anomalies introduced by stegware.

Feature extraction techniques aim to highlight variations or artifacts caused by the presence of hidden data, making it easier to distinguish between cover and stego media. The relevance of feature engineering in steganalysis is crucial. By selecting and extracting informative features, it enables the construction of robust models that can effectively detect and analyze stegware. The choice of features significantly influences the accuracy, efficiency, and interpretability of steganalysis algorithms. Well-designed features that capture unique characteristics introduced by stegware can provide reliable cues for identifying hidden information, while also facilitating the analysis of steganographic

Table 1. Feature Engineering Techniques

Feature Engineering Approach	Specific Technique	Description	Advantages
Statistical Analysis	Entropy	Measures the randomness of pixel values, highlighting deviations caused by steganography	Simple and effective
Texture Analysis	Co-occurrence Matrices	Captures spatial relationships between pixel values, hidden textural patterns	Sensitive to textural alterations
Frequency Domain Analysis	Discrete Wavelet Transform (DWT)	Analysis frequency components, detecting modifications in different scales and directions	Can reveal specific stenographic methods
Deep Learning Features	Convolutional Neural Networks	Extracts hierarchical features, recognizing complex stegware patterns through learned representations	Can learn complex patterns
Statistical Analysis	Entropy	Measures the randomness of pixel values, highlighting deviations caused by steganography	Simple and effective

techniques. Moreover, feature engineering helps in addressing the challenges posed by stegware, such as the ability of stegware to conceal information with minimal visual or auditory impact. By focusing on relevant features [3], steganalysis algorithms can effectively differentiate between cover media and stego media, even in scenarios where stegware introduces subtle modifications.

3.1 Feature Selection Techniques

Feature selection plays a vital role in stegware analysis as it helps identify the most relevant features that contribute to the detection [1] of hidden information. Various feature selection methods are utilized in steganalysis to reduce dimensionality and enhance the performance of steganalysis algorithms. The overview of feature selection techniques in stegware [4, 5] (Fig. 1).

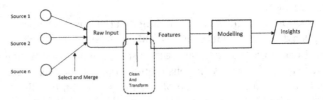

Fig. 1. Feature Selection method

- **Filter Methods:** Filter method assess the relevant features in view with their intrnsic characteristics, independent of any specific ML learning algorithm. These methods analyze statistical measures, information-theoretic metrics, or correlations between features and the target variable (cover or stego). In Statistical Measures the Features can be ranked based on their statistical properties, such as mean, standard deviation, skewness, or kurtosis. Higher-order statistical moments can capture the presence of hidden data by detecting deviations from the expected distributions. The Information-Theoretic Metrics measures the mutual information or entropy quantifies the information of the feature related to target variable. The Features with huge information gain or mutual information tend to be relevant for steganalysis. Correlation Analysis measures the Pearson's correlation coefficient and assesses the linear relationship between features and the target variable. Features with high correlation or anti-correlation to the target are selected.

- **Wrapper Methods:** Wrapper methods are used in evaluating the performance of different feature subsets by incorporating machine learning algorithms. These methods create subsets of features and train the classifier to assess the predictive power of each subset. Sequential Forward Selection (SFS) starts with a null feature set and iteratively adds a feature that improves the performance of the classifier. It follows until the desired level of performance is achieved or no further improvement is observed. Sequential Backward Selection (SBS) starts with all features and eliminates them one by one based on their impact on classifier performance. Features with the least contribution are removed until the desired performance level is attained.

- **Embedded Methods:** Embedded methods uses feature selection within the learning algorithm itself (Table 2).

Table 2. Feature Selection Techniques

Feature Selection Techniques	Types/Techniques	Uses
Filter Methods	-Statistical Measures Information -Theoretic Metrics Correlation Analysis	Preprocessing step in steganalysis Dimensionality reduction
Wrapper Methods	-Sequential Forward Selection -Sequential Backward Selection	-Performance optimization in steganalysis -Subset selection for model training
Embedded Methods	– Regularization Techniques –Tree-Based Feature Importance	-Integrating feature selection into model training -Enhancing model performance in steganalysis

These methods optimize both feature selection and model performance simultaneously. Regularization methods, such as Lasso (L1 regularization) or Ridge regression (L2 regularization), penalize the model for non-informative features, leading to automatic feature selection during training.

Specific feature selection techniques are often applied in steganalysis based on the characteristics of stegware and the dataset under investigation. Each technique has its advantages and limitations, including computational complexity, robustness to noise, and potential for over fitting. The selection of an appropriate feature selection technique should be based on the need and constraints of the steganalysis.

3.2 Feature Extraction Techniques

Feature extraction techniques play a crucial role in stegware analysis as they help capture relevant information from digital media that can be used for steganalysis. Various feature extraction methods are employed in steganalysis to extract discriminative features that distinguish between cover and stego [6, 7].

- **Statistical Feature Extraction Methods:** Statistical feature extraction methods analyze the statistical properties of digital media to extract relevant features. Computation of Statistical Moments possesses features like mean, standard deviation, skewness, or kurtosis that are computed to capture statistical characteristics of pixel intensities or color channel values. Higher-order statistical moments can reveal deviations from expected distributions, indicating the presence of hidden data. Pixel Intensity Distributions has Histograms or probability density functions of pixel intensities that provide insights into the distribution patterns. Features such as entropy, contrast, or energy can be derived from these distributions to quantify image characteristics. Color Channel Statistics: works on the Statistical properties of color channels, such as mean, standard deviation, or correlation coefficients, are extracted to capture color-related information. Color channel statistics can be computed in different color spaces, such as RGB, HSV or YUV.

- **Transform-Based Feature Extraction Techniques:** Transform-based feature extraction methods use mathematical transforms to analyze the frequency or spatial characteristics of digital media. Discrete Cosine Transform (DCT) a very popular steganalysis method to convert spatial image file into the frequency domain. DCT coefficients can capture energy compaction and high-frequency details, which are often affected by steganographic embedding. Wavelet Transform decomposes images into multiple frequency sub bands, capturing both global and local frequency characteristics. Features such as wavelet coefficients, energy distribution, or sub band statistics can be derived for steganalysis. Fourier Transform analyzes the frequency components of an image. Features such as magnitude spectra, phase information, or frequency histograms can be extracted for steganalysis.

- **Spatial Domain Feature Extraction Methods:** Spatial domain feature extraction methods analyze the spatial characteristics of digital media, focusing on patterns, textures, or edges. Texture Analysis Features are related to texture patterns, such as local binary patterns (LBP), co-occurrence matrices, or Gabor filters, can be computed to capture textural information. Texture features are useful in detecting steganographic changes affecting local image regions. Edge-based features, such as edge density, edge orientation histograms, or gradient magnitude, can be extracted to capture edge-related information. Changes in edges caused by steganographic embedding can be identified through these features are focused on Edge detections. These features are robust to geometric and photometric transformations and can assist in steganalysis [8]. Specific feature extraction techniques are employed in steganalysis based on the characteristics of stegware and the analysis goals. Each technique has its advantages and limitations, including computational complexity, sensitivity to noise, and ability to capture specific steganographic changes (Fig. 2).

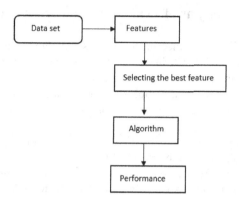

Fig. 2. Feature Selection Process

Table 3. Comparison of Feature extraction techniques

Feature Extraction Method	Key Characteristics	Common Features Extracted	Typical Dataset(s) Used	Evaluation Metric(s)	Computational Complexity
Statistical Feature Extraction Methods	Captures statistical properties of data	Mean, Variance, Skewness, Kurtosis, Entropy, etc	Image, audio, text steganography datasets	Detection Rate, False Positive Rate	Low to Moderate
Transform-Based Feature Extraction Techniques	Analyzes data in different domains (e.g., frequency, spatial)	Discrete Cosine Transform (DCT) Coefficients, Wavelet Coefficients, FFT Components, etc	Image, audio, video steganography datasets	Accuracy, F1 Score	Moderate to High
Spatial Domain Feature Extraction Methods	Focuses on characteristics of data in its original spatial domain	Texture Features (Local Binary Patterns, Gabor Filters, etc.), Edge Features, Color Histos	Image, texture analysis datasets	True Positive Rate, False Negative Rate	Moderate to High

Table 3 provides a comprehensive view of the feature extraction techniques on stegware. This table showcases the key characteristics of the method, the possible features that can be extracted and the datasets that works well on this process [17].

4   Comparative Analysis of Feature Engineering Techniques

In stegware analysis, feature engineering techniques, including feature selection and feature extraction [9], plays a vital role in enhancing the accuracy and efficiency of steganalysis algorithms. It is important to compare and analyze these techniques to understand their effectiveness, computational complexity, and impact on the overall performance of steganalysis. Here is a comparative analysis of feature engineering techniques in stegware (Table 4):

The dataset [10, 11], which consists of 10000 Gy scale images is used for applying the feature selection and extraction process (Fig. 3).

Here, a sample data is from the dataset is used to show the feature selection methods on the image. The results are displayed with histograms to understand the difference. The feature selection techniques like color corrector, contrast adjustment, edge corrector are used. From the dataset, a sample image is taken and Feature Extraction techniques

Table 4. Comparison between Feature Selection and Extraction Methods

Topic	Feature Selection	Feature Extraction
Definition	Selecting a subset of relevant features from the dataset	Transforming the original features into a new feature space
Approach	Selecting the most informative features based on relevance	Creating new features by combining or transforming existing ones
Purpose	Reduce dimensionality and improve model interpretability	Represent the data in a more informative or compact manner
Information Preservation	Removes irrelevant or redundant features	Preserves the original information by creating new features
Techniques	Univariate selection, Recursive Feature Elimination (RFE), L1 Regularization (Lasso), Mutual Information	Principal Component Analysis (PCA), Linear Discriminant Analysis (LDA), Non-negative Matrix Factorization (NMF), AutoEncoders
Supervised/ Unsupervised	Can be used in both supervised and unsupervised settings	Often used in unsupervised settings but can be adapted for supervised learning
Feature Subset	Selects a subset of existing features from the original dataset	Creates a new set of features based on the original dataset
Interpretability	Preserves the interpretability of selected features	May result in less interpretable features
Computational Complexity	Generally less computationally expensive compared to feature extraction methods	Can be computationally expensive, especially for complex transformations
Data Requirement	Requires labeled data or target variable information	Can be used with labeled or unlabeled data
Algorithm Compatibility	Compatible with a wide range of machine learning algorithms	Requires specific algorithms designed for the transformed feature space

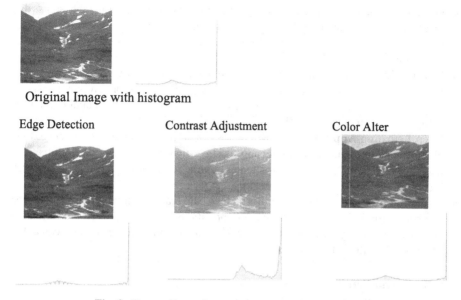

Original Image with histogram

Edge Detection   Contrast Adjustment   Color Alter

Fig. 3. Feature Extraction techniques on Bossbase dataset

like Edge Detection, Contrast Adjustment, and Color Alter. To show the variations in the feature extraction methods, histogram is generated on the original image and the modified images as well. Changes in the histogram makes evident on the different extraction techniques. Thus, based on the requirements the techniques may vary.

5 Challenges and Future Directions

Feature engineering in stegware analysis is a challenging and evolving field that presents several research questions and opportunities for future advancements. Here are some key challenges, open research questions, and potential directions for future research and development in feature engineering [12] for stegware.

5.1 Challenges

- **Adaptive Steganalysis:** As steganographic techniques evolve and become more sophisticated, feature engineering needs to adapt to detect hidden information effectively. Developing adaptive feature engineering techniques that can capture and analyze emerging steganographic methods is a challenge.
- **High-Dimensional Data:** Steganalysis often deals with high-dimensional data, such as images or videos, which pose challenges in feature selection and extraction. Handling and processing high-dimensional data efficiently is an ongoing challenge.
- **Robustness to Attacks:** Steganographic techniques can be designed to evade steganalysis algorithms. Feature engineering methods need to be robust and resistant to evasion attacks to ensure accurate detection of hidden data.

- **Generalizability:** Feature engineering techniques should be generalizable across different steganographic algorithms, file formats, and media types. Ensuring the effectiveness and robustness of feature engineering methods in diverse stegware scenarios is a challenge.
- **Feature Fusion:** How can different types of features be effectively fused to improve the detection performance in steganalysis? Exploring methods for combining features from different domains, such as statistical, transform-based, and spatial domain, is an open research question [14].
- **Deep Learning Approaches:** How can deep learning techniques be leveraged for automatic feature extraction in steganalysis? Investigating deep learning architectures and transfer learning strategies for feature extraction can be a promising research direction.
- **Explainability and Interpretability:** How can feature engineering methods be designed to provide explanations and interpretations of steganalysis results? Enhancing the transparency and interpretability of feature engineering techniques can aid in understanding the detection process.

5.2 Scope and Future Directions

The scope of the work encompasses an overview of the Feature engineering techniques relevant to Steganalysis. This study focuses on identifying the most relevant features in the images which aims to distinguish between the stego (Stealthy) and the cover medium (clean) images. Several approaches are covered, with a comparison as well. When the best feature engineering technique is chosen, it results in the effectiveness and accuracy in detecting the stegware. A combination of the technique works fine at most of the areas to gain a higher detection of the stegware.

- **Hybrid Feature Engineering:** Exploring hybrid feature engineering approaches that combine the strengths of different feature selection and extraction techniques can lead to improved performance in steganalysis. Integrating statistical, transform-based, and spatial domain features can enhance the detection capabilities.
- **Domain-Specific Feature Engineering:** Developing feature engineering techniques tailored to specific stegware domains, such as image steganography, audio steganography, or network steganography, can improve detection accuracy and efficiency. Customized feature engineering methods can exploit domain-specific characteristics [15, 16].
- **Machine Learning for Feature Engineering:** Leveraging machine learning algorithms for automated feature engineering can enhance the efficiency and effectiveness of steganalysis. Investigating approaches such as genetic algorithms, reinforcement learning, or automatic feature selection methods can streamline the feature engineering process.

Thus, feature engineering in stegware analysis faces challenges related to adaptive steganalysis, high-dimensional data, robustness to attacks, and generalizability. Future research should focus on feature fusion, deep learning approaches, explainability, and interpretability.

6 Conclusion

In conclusion, this survey paper has provided a comprehensive overview of feature engineering in stegware analysis. We have emphasized the significance of feature engineering in steganalysis and its role in detecting and analyzing hidden data within digital media. The survey has covered several selection and extraction methods along with their comparisons in each base, highlighting their effectiveness and computational complexity. Additionally, we have discussed feature extraction techniques such as statistical feature extraction, transform-based feature extraction, and spatial domain feature extraction, and their ability to capture steganographic changes. Through a comparative analysis, we have examined the strengths, limitations, and impact of different feature engineering choices on steganalysis algorithms. The paper has also identified challenges, open research questions, and potential future directions, including hybrid feature engineering, domain-specific techniques, and the overview deep learning and attention mechanisms. Thus the future of feature engineering in stegware holds great potential for advancing steganalysis techniques and improving the detection and analysis of stegware.

However, the best feature engineering techniques for the stegware analysis may completely depend upon various factors like the type of stegware, the model, and goal of the analysis, resources available and the nature of the digital medium.

References

1. Monika, A., Eswari, R.: An Ensemble-based Stegware Detection System for Information Hiding Malware Attacks, 06 November 2022, PREPRINT (Version 1) available at Research Square [https://doi.org/10.21203/rs.3.rs-613020/v1]
2. Boeschoten, S., Catal, C., Tekinerdogan, B., Lommen, A., Blokland, M.: The automation of the development of classification models and improvement of model quality using feature engineering techniques. Expert Syst. Appl. **213**, 118912 (2022). https://doi.org/10.1016/j.eswa.2022.118912
3. Gibert, D., Planes, J., Mateu, C., Le, Q.: Fusing feature engineering and deep learning: a case study for malware classification. Expert Syst. Appl. **207**, 117957 (2022). https://doi.org/10.1016/j.eswa.2022.117957
4. Miao, J., Niu, L.: A survey on feature selection. Procedia Comput. Sci. **91**, 919–926 (2016). https://doi.org/10.1016/j.procs.2016.07.111
5. Panda, M., Abd Allah, Al.M., Hassanien, A.E.: Developing an efficient feature engineering and machine learning model for detecting IoT-botnet cyber attacks. IEEE Access **9**, 91038–91052 (2021). https://doi.org/10.1109/ACCESS.2021.3092054
6. Daeef, A.Y., Al-Naji, A., Nahar, A.K., Chahl, J.: Features Engineering to Differentiate between Malware and Legitimate Software. Appl. Sci. **13**(3), 1972 (2023). https://doi.org/10.3390/app13031972
7. Salau, A.O., Jain, S.: Feature extraction: a survey of the types, techniques, applications. In: 2019 International Conference on Signal Processing and Communication (ICSC), NOIDA, India, pp. 158–164 (2019). https://doi.org/10.1109/ICSC45622.2019.8938371
8. Salau, A., Jain, S.: Feature extraction: a survey of the types. Tech. Appl. 158-164 (2019). https://doi.org/10.1109/ICSC45622.2019.8938371
9. Cao, W., Li, S., Li, W., Li, J.: A novel feature extraction method based on deep learning for steganalysis. Multimedia Tools Appl. **77**(16), 20905–20921 (2018)

10. Xie, G., Ren, J., Marshall, S., Zhao, H., Li, R., Chen, R.: Self-attention enhanced deep residual network for spatial image steganalysis. Digit. Signal Process. **139**, 104063 (2023). https://doi.org/10.1016/j.dsp.2023.104063

11. Altae, Aymen A and Rad, Abdolvahab Ehsani and Tati, Reyhaneh, Comparative Study on Effective Feature Selection Methods (February 13, 2023). International Journal for Innovative Engineering & Management Research, Forthcoming, Available at SSRN: https://ssrn.com/abstract=4357775

12. Liu, D.R., Li, H.L., Wang, D.: Feature selection and feature learning for high-dimensional batch reinforcement learning: a survey. Int. J. Autom. Comput. **12**(3), 229–242 (2015). https://doi.org/10.1007/s11633-015-0893-y

13. Uddin, M.F., Lee, J., Rizvi, S., Hamada, S.: Proposing enhanced feature engineering and a selection model for machine learning processes. Appl. Sci. **8**(4), 646 (2018). https://doi.org/10.3390/app8040646

14. Smith, J.K., Johnson, A.B.: An empirical analysis of feature engineering for predictive modeling. J. Data Sci. **15**(2), 123–145 (2018)

15. Mays, Mitchell & Drabinsky, Noah & Brandle, Stefan. (2017). Feature Selection for Malware Classification

16. Yadav, N., Singh, M.: A novel approach for feature extraction using CNN for steganalysis. Multimedia Tools Appl. **80**(21), 31225–31247 (2021)

17. Babu, J., Rangu, S., Manogna, P.: A survey on different feature extraction and classification techniques used in image steganalysis. J. Inf. Secur.Secur. **08**, 186–202 (2017). https://doi.org/10.4236/jis.2017.83013

Text Summarization Using Deep Learning: An Empirical Analysis of Various Algorithms

Namita Kiran[✉], Leena Ragha, and Tushar Ghorpade

Department of Computer Engineering, Ramrao Adik Institute of Technology,
D. Y. Patil Deemed to be University Nerul, Navi Mumbai, India
namitakiran20@gmail.com, {leena.ragha,
tushar.ghorpade}@rait.ac.in

Abstract. Text Summarization is rephrasing the text into a shorter, concise form while preserving its original meaning. Various researchers have worked on this domain using deep learning techniques, but still there is a scope to produce a concise and meaningful summary. In this work, we perform empirical study of neural models like seq2seq and Transformers. The focus is on architectural advancements and their impact on the summarization. Also, we discuss the introduction of the Attention module and its impact on the performance of the algorithm. Through these experiments, a comparison is drawn among various models. A reference sentence is taken and various algorithms are applied on the sentence. The dataset used to train the model is amazon food review. The generated summary using each algorithm is evaluated on ROUGE score and conclusion is drawn based on results obtained. Although Transformer outperforms seq2seq models however, seq2seq can be used for certain tasks. Also, the significance of attention module on summary can be visualized through results.

Keywords: Seq2seq · attention · transformer and multi-head Transformer

1 Introduction

Text Summarization [3, 4] is encapsulating the texts from one or more sources into one that is precise, informative and half the size of original texts. Reading a long document could be time-consuming and exhaustive, sometimes unnecessary. Also, human understanding may be error-prone. Text summarization presents the gist of the document. Significant words and its context are preserved so that the original meaning of the paragraph is not altered. Useful information is extracted and presented to the reader. It is used in various applications namely, Research analysis, News summary, Abstract generation, Product review, Email summary, Students answer sheet evaluation, etc. There are various approaches followed by the researchers to produce text summarization. They are broadly categorized as follows.

- Extractive summarization [6]-Important sentences are extracted "as it is" from the original document.

© The Author(s), under exclusive license to Springer Nature Switzerland AG 2024
S. Satheeskumaran et al. (Eds.): ICICSD 2023, CCIS 2122, pp. 175–185, 2024.
https://doi.org/10.1007/978-3-031-61298-5_14

- Abstractive summarization [6]-This technique rewrites the paragraph into shorter, concise form capturing the salient features of original text.

Hence, Abstractive summarization techniques are very complex as they need to reform the new sentences based on the important information components. Both types of summarization is required depending on task domain. In tasks like students answer sheet summarization, research analysis, etc. extractive summarisation may be needed as they are critical tasks and loss of important information or generation of incorrect summary may prove costly. For sentiment analysis, product review, News summary and other informal tasks abstractive summarisation can be used. We propose to look into both the categories of deep learning techniques to analyze the impact. The work compares various deep learning algorithms based on their architecture and tries to understand the limitations of each of them. Also it studies the architectural advancements that has taken place and its impact on summary generation using a test data. The summary obtained is empirically compared using ROUGE(Recall-Oriented Understanding for Gisting Evaluation) metrics. This study further compares the result, obtained using the above models on the same test sentence and discusses which model is best suited to what type of text summarization tasks. It brings out the limitations of each of them and also what type of summarization(extractive or abstractive) can be obtained using them. To get a meaningful summary, it is advised that the paragraph should be well-written and have a theme or context. Poorly written sentences may generate random results. Human generated summary may have grammtical errors. Use of transformer minimizes that.

The paper is organized with Sect. 2 concentrating on the literature survey of the research work in both extractive and abstractive domains. Based on the survey analysis, we chose 3 architectures explained in Sect. 3 for the empirical study. Section 4 discusses the results and finally the conclusion and the references are presented.

2 Related Works

The detailed survey is carried out in both extractive and abstractive domains as discussed in the following subsections to understand the various architectures, their complexity and impact on the summary. The paper "Attention Is All You Need" [1] released by Google proposed a new architecture model-Transformer that used an attention mechanism. The use of the transformer entirely removed dependence on recurrent and convolutional networks. Studies on machine translation tasks showed that they can be significantly parallelized and need comparatively less time to train. The results were superior in quality. Google AI in 2019 published a paper named "BERT: Pre-Training of Deep Bidirectional Transformers for Language Understandings" [5]. It introduced a Bidirectional Encoder Representation Transformer(BERT), a deep pretrained bidirectional encoder-only model. The representation of a word is dependent on both the left and right context of the word. With addition of just one output layer, BERT can be fine-tuned to create pioneering innovation. Without any major changes to architecture, BERT models are used for an extensive number of applications. Authors Dong Qiu and Bing Yang in their research paper titled "Text Summarization based on multi-head self-attention mechanism and pointer network" [8] introduced an improved version of multi-headed self-attention in the coding phase of the experiment. In the training stage the model gives higher weight to the semantics and summary information hence the generated summary is more exact and coherent. The pointer network model was used in the experiment and to tackle the problems of repetitive and out-of-vocabulary words. Coverage mechanism was improved. Compared to other models, the generated abstract using multi-headed transformer with self-attention and pointer network showed better quality results. "Learning to summarize from human feedback" [12] by Open AI adopts a novel approach to fine-tune a summary to improve its quality. In previous papers, the quality of summary is categorized as good on the basis of similarity to a reference summary and its ROUGE score the authors further fine-tuned this process by collecting a substantial amount of high-quality dataset of summaries that has been allotted preferences based on human feedback. The model is further trained to match human preferences. Summary is optimized based on feedback.

The above studies present architecture of various models and its use for text summarization task. Based on this survey, we find that Transformer models are very promising in almost all kinds of NLP tasks. However, this paper shows that seq2seq models using RNN as its base is efficient to be used in certain tasks as presented in Sect. 4. We further explore them in detail w.r.t architecture and try to find the strengths and limitations of each of them.

3   Model Architecture

Summary can be generated using various deep learning methods. This work analyzes two approaches - Sequence-2-sequence and Transformer.

3.1   Seq2seq Model

Sequence to Sequence or commonly referred as seq2seq [7] is a neural network model introduced by Google in 2014. It is a many-many model, where several input words are fed and output is a sequence of words shorter in length. The model is used in sequence based problems especially in the cases where inputs and outputs are of different lengths as in the case of text summarization. The architecture of the seq2seq model can be broadly distinguished into two parts as shown in Fig. 1.

- Encoder-It takes the source document as input and creates an internal representation. It can be said as sequence-to-Vector as it takes a sequence of inputs and converts it into an intermediate representation or a context vector. The context vector incorporates the essence of input.
- Decoder-The intermediate output generated from the encoder is the input to the decoder. Decoder feeds the input vector multiple times at each time step and produces an output.

The encoder-decoder may use plain RNN, LSTM-RNN or GRU as its component. Plain RNN is now rarely used because of vanishing gradient problems.

Se2seq with LSTM. LSTM(Long Short Term Memory) [9,13] RNN is used in the encoder-decoder. Encoder reads the entire input sequence, one word at each timestep. The input is processed at every timestep and contextual information is captured. LSTM encoder produces two outputs- a hidden state(h) and a cell state(c). h and c forms the internal state. This internal state is input to the decoder. LSTM architecture consists of three gates-input, output and a forget gate that provide LSTM long term dependency. In Fig. 1 the hidden state(h_0) and cell state(c_0), initialized to zero vectors are inputs to the encoder. h_4 and c_4 are the output vectors from encoder.

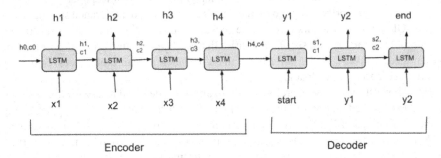

Fig. 1. Seq2seq with LSTM encoder-decoder [10].

The final hidden(h_4) and cell(c_4) states from the encoder is input to the decoder. < start > and < end > tokens indicate start and ending of sequence. This is necessary because in decoder the length of the target sequence is uncertain while we decode the test sequence. y_1 and y_2 are the predicted outputs s_1 and s_2 are hidden states whereas c_1 and c_2 are cell states.

The use of LSTM-RNN outperforms traditional RNNs. First it deals with the problem of vanishing gradient. The three gates help in determining which information the cell should remember and what it should forget. It works great with sequence related tasks. However, it is not sufficient. LSTMs cannot be trained in parallel. It takes more memory and time during the training phase. Models easily overfit. Also, due to inconsistent weight initialization, dropout is more difficult.

Seq2seq with Attention. Seq2seq with Attention [1] consists of, in addition to encoder and decoder an attention layer. Attention layer deals with the problem of long sequences. The encoder compresses the entire input and creates a context vector. Attention layer focuses on only important words in a sentence rather than the entire sequence. It is an attempt to focus on only a few relevant words in a given sentence and eliminate others. So the key concept in the attention technique is that instead of looking at all the words in the source document, only specific parts of the source sequence are considered for the resultant sequence. The context vector is more expressive in nature filtered via an attention layer, rather than being a fixed-length encoding as in the simple Encoder-Decoder architecture. After text preprocessing input is fed to encoders into the RNN cells. RNN may be plain or any variant like LSTM, Bi-LSTM or GRU. The output from each encoder is the input to a feed forward neural network. Past studies suggest 40–50 is the optimal number of inputs to the FFN. The Neural Network undergoes training to perfect its weight. Figure 2 shows $X_1, X_2 \ldots X_T$ is the input vector to the bidirectional

LSTM encoder. For each input sequence a series of annotations $h_1, h_2 \ldots h_T$ is created.

$$h_j = \left[h_j^T ; h_j^T \right]^T \tag{1}$$

These annotations are nothing but a concatenation of forward and backward hidden states in the encoder. Simply put, $h_1, h_2, \ldots h_T$ vectors are the representation of input words.

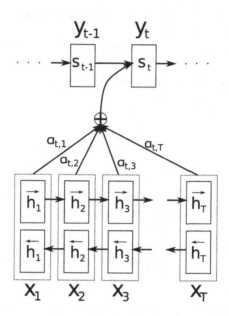

Fig. 2. Block diagram of seq2seq with Attention Layer [1]

Attention Layer consists of

- Context vector
- Attention weight
- Alignment score

In normal encoder-decoder only the last hidden state i.e. h_T would be considered for output generation, but in the attention mechanism all the hidden states are used to create a context vector.

$$c_i = \sum_{j=1}^{T_x} \alpha_{ij} h_j \tag{2}$$

The context vector c_i is the weighted sum of all the encoder hidden states $h_1, h_2 \ldots h_T$ and attention weights. The context vector is used to calculate the final output from the decoder. The above formula calculates the context vector(c_i).

$$a_{ij} = \frac{exp(e_{ij})}{\sum_{k=1}^{T_x} exp(e_{ik})} \tag{3}$$

In the above equation, α_{ij} is the weight of h_j calculated by the softmax activation function. Higher the weight of input, higher will be its influence on output sequence.

$$e_{ij} = a(s_{i-1}h_j) \tag{4}$$

e_{ij} is the alignment score of a feedforward neural network with a as attention function that tries to realize the alignment between input j and output i. It is based on previous decoders hidden state s_{i-1} and current hidden state, h_j. The decoder decides which part of the sequence it should pay more attention to rather than treating all inputs equally. Even with the addition of the attention layer seq2seq provides only local dependency. They process data sequentially therefore, computation is slow. This makes it difficult to train on large amounts of data. Summaries generated using seq2seq are abstractive in nature. They may be incoherent. Also summaries generated may be logically incorrect. 2–3 lines of input will generate 2–3 words of summary.

3.2 Transformers

Transformers [2] is a neural network architecture developed by Google and UoT in 2017. Original transformer has six sets of encoder-decoders as its core component and uses a modified version of attention called self-attention. The output from *layer l-1* goes to higher layer *l*. It has an advantage over sequential models as it avoids using the principle of recurrence and focuses on attention mechanisms enabling global dependency rather than local. It is a preferred model to today's NLP challenges and can be used for a wide range of language modeling tasks. Transformers allow for parallelization and hence can train on huge corpus of data which is difficult with sequential models.They can attain high summarization quality even when trained for less amount of time. Further, the concept of self-attention provides a more improved quality of results as it helps to focus only on relevant parts of a sentence.

Original Transformers consist of a stack of encoders fully connected to a stack of decoders shown in Fig. 3. Input sequence is fed into encoders. The first encoder gets word embedding as the input. Each of the word embedding flows through the two layers of encoder i.e. self-attention and feed-forward.

The first encoder receives the input and starts processing it by passing it first to the self-attention module and then through the feed-forward neural network as shown in Fig. 4. The output from feed-forward goes to the encoder just above it. The Transformer encoder consists of self-attention, a modified version of attention. The Self-Attention module takes n inputs and gives n outputs. It allows interaction among inputs(self) to decide which input it should pay more attention to. The output is the aggregate of this interaction and alignment score. Self-attention layer contextually encodes the input sequence. It helps a sentence to understand the correct meaning of the word relative to the sentence or the paragraph it is used in.

As shown in Fig. 5, the sentence "I went to the bank to withdraw some money" each word is first processed into a vector. Each vector forms Key. The Query is an individual word whose relationship with all other words including itself is to be found using cosine similarity. Dot product between Keys (I,went,to, the,bank,to,withdraw, some,money) and Query(I) tells the relationship of word "I" with each input word(v1,v2…vn). Similarly,

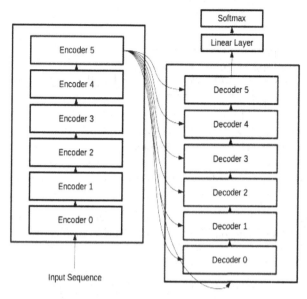

Fig. 3. Block diagram of transformer [1]

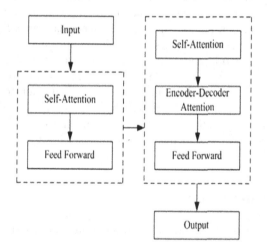

Fig. 4. Decomposition of encoder-decoder into self-attention and Feed-forward network [2]

relation among all words is calculated using cosine similarity to obtain the score. The scores are normalized and then multiplied to Value vectors of n1 matrix which is an input to FFN. However, the feed-forward layer is not capable of processing eight matrices. It expects a single matrix. The matrices Query(Q), Key(K) and Value(V) are obtained for each word embedding(w) of 512 dimensions.

$$S = QK^{\mathrm{T}} \tag{5}$$

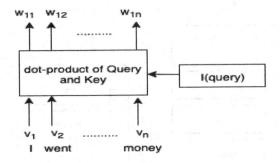

Fig. 5. Illustrative example

The score matrix(S) is obtained by dot product of Query (Q) and transpose of Key(K). The key(dk) is of 64 dimensions. The attention score is divided by the square root of dk..

$$Attention(Q, K, V) = softmax_k \left(\frac{QK^T}{\sqrt{d_k}} \right) V \tag{6}$$

They are then adjusted via normalization. Each value vector gets multiplied by its softmax score. The intention is to preserve the value of the word that is to be focussed and draw-out irrelevant words by multiplying them with very small values like 0.001. The key feature is modeling global dependencies by learning relationships between elements of inputs through a self-attention mechanism by pairwise correlation. Recent developments have led to Multi-head Attention [8]. In Multi-head attention there are multiple sets of key, value, Query as opposed to one in single-head attention transformers. Input embedding is multiplied to weight matrices $Q_w/K_w/V_w$ to obtain a Q/K/V matrix. Eight different self-attention calculations are done with different weight matrices resulting in eight different z matrices. The eight matrices are concatenated to obtain z. Transformers produce extractive as well as abstractive summarization.

4   Experiment and Result

The dataset used is amazon food review. Test results are collected from following:-

1. Encoder-Decoder with LSTM as its base- The review and summary column in the given dataset undergoes text preprocessing. The dataset is split in training and testing. Three encoders and three decoders are used. Optimizer and loss function used is RMSprop and sparse categorical crossentropy res.
2. Encoder-Decoder with attention-An additional attention.py file added is added to the outputs of encoder and decoder. It also uses LSTM-RNN.
3. Transformer- T5(Text-to-Text Transfer Transformer) [15] with muti-headed attention is used. It can be replaced with BERT or GPT that generates similar results.

The three broad architectures have been compared in Table 1 based on summary evaluation by Recall-Oriented Understanding for Gisting Evaluation(ROUGE) score.

$$ROUGE\ 1 = \frac{no.\ of\ common\ unigrams\ in\ C\ and\ R}{no.\ of\ unigrams\ in\ R} \tag{7}$$

where C is candidate and R is reference summary. Reference summary is a human generated summary that is compared to a system generated summary. The Table 1 shows the generated summary of sample sentence using seq2seq, seq2seq with attention and Transformer. The precision, recall and f1measure metrics corresponding to ROUGE1 is obtained. Although the seq2seq model correctly expresses the sentiment of text, all the three metrics are 0 because of the absence of any common unigram between C and R. It specifies that the sentence expresses positive sentiment about something, but rest of the information is lost because of limited memory. It is advisable to be used with 2–3 lines of sentences containing 25–50 words. Seq2seq with attention contains two common unigrams and expresses sentiment of article correctly. Also, it tells that the information is about a 'product' but other details are not retained. It can be useful for 3–5 lines of text ranging from 75–100 words. The summary using seq2seq may be sometimes incorrect and is unreliable. Use of attention module improves the result upto some extent. Further, during text preprocessing stage stopwords are filtered out. Hence, the result may not be grammatically coherent. Let us analyse another sample text.

Sample text- *"This seems a little more wholesome than some of the supermarket brands, but it is somewhat mushy and doesn't have quite as much flavor either."*

Summary-

LSTM without attention-*great*

LSTM with attention-*great taste*

The above example demostrate an error case where seq2seq fails to predict correctlty. The sentiment of the review is negative whereas summary generated is positive.

Transformer model outperform deep learning methods in all the three parameters. The summary is informative and grammtically correct. It can generate abstractive as well as extractive summarization as per users need. It works well with few lines of text or long articles consisting of more than 1000 words. However, it is better to be used with long text. Also it takes care of word case unlike the above two that converts all words into lowercase while text preprocessing. However, Transformers require huge computational resources which may not be available and also not required. So, for short informal texts used for applications like product review, sentiment analysis deep learning method is beneficial and for long documents with critical applications, transformer model works best.

Table 1. Empirical analysis of models

Sample text = *"I am very satisfied with my Twizzler purchase. I shared these with others and we have all enjoyed them. I will definitely be ordering more"*		**Reference summary =** *"very satisfied with Twizzler purchase. Shared product with others. Ordering more."*	
Model	Summary (type)	Review of summary	Metrics
Seq2seq	great (abstractive)	Works for short text. Loses context or incorrect prediction if text size increases Good for product review or sentiment analysis Training is slow because of sequential input /output	precision $= 0$ recall $= 0$ fmeasure $= 0$
Seq2seq with Attention Layer	great product (abstractive)	Works for short text. Loses context as length increases Accuracy is better than plain seq2seq. Reatains more information Training is slow as input/output is sequential	precision $= 0.09$ recall $= 0.5$ fmeasure $= 0.15$
Transformer	i am very satisfied with my Twizzler purchase (extractive)	Suitable for long and multi-document text.. Summary is accurate and reliable. Can be used for critical tasks like research analysis, answer- sheet evaluation. Computation is faster. Produces both extractive and abstractive summarization	precision $= 0.45$ recall $= 0.63$ fmeasure $= 0.53$

5 Conclusion

Text processing has evolved in recent years. Human generated summary may be error-prone, Transformers help to overcome those. These are the state-of-the-art algorithms giving efficient and human-like results. The Multi-head Attention layer has given a whole new dimension to text processing. Many pre-trained models are available that saves our computation time and resources. Various issues to text processing have been resolved, however bigger challenges still persist. The most common challenge is increased architectural complexity. To train such complex models, a huge dataset is required, which may not be always available. Also, the definition of a good summary differs from person to person so rating the quality of summary is subjective and varies depending on readers interests and what it expects in the summary. In terms of accuracy, transformers provide good accuracy. However, still improvements are needed before it is impossible to differentiate between machine and human translation. Also, existing models have been trained on English language, not much work has been done in regional languages.

References

1. Vaswani, A., et al.: Attention is all you need. In: Advances in Neural Information Processing Systems. arXiv preprint arXiv:1706.03762 (2017)
2. Chitty-Venkata, K.T., Emani, M., Vishwanath, V., Somani, A.K.: Neural architecture search for transformers: a survey. IEEE Access (2022)
3. Boorugu, R., Ramesh, G.: A survey on NLP based text summarization for summarizing product reviews. In: 2020 Second International Conference on Inventive Research in Computing Applications (ICIRCA), pp. 352–356 (2020)
4. Elsaid, A., Mohammed, A., Ibrahim, L.F., Sakre, M.M.: A comprehensive review of Arabic text summarization. IEEE Access **10** (2022)
5. Devlin, J., et al.: BERT: pre-training of deep bidirectional transformers for language understanding. arXiv preprint arXiv:1810.04805 (2018)
6. Mridha, M.F., Lima, A.A., Nur, K., Das, S.C., Hasan, M., Kabir, M.M.: A survey of automatic text summarization: progress, process and challenges. IEEE Access **9**, 156043–156070 (2021)
7. Shi, T., Keneshloo, Y., Ramakrishnan, N., Reddy, C.K.: Neural abstractive text summarization with sequence-to-sequence models. ACM/IMS Trans. Data Sci. **2**(1), 1–37 (2021)
8. Qiu, D., Yang, B.: Text summarization based on multi-head self-attention mechanism and pointer network. Complex Intell. Syst. **8**, 555–567 (2022)
9. Shaik, T., et al.: A review of the trends and challenges in adopting natural language processing methods for education feedback analysis. IEEE Access **10**, 56720–56739 (2022)
10. Divya, K., Sneha, K., Sowmya, B., Rao, G.S.: Text Summarization using Deep Learning. Int. Res. J. Eng. Technol. **7**(05), 3673–3677 (2020)
11. Burmani, N., Alami, H., Lafkiar, S., Zouitni, M., Taleb, M., Nahnahi, N.E.: Graph based method for Arabic text summarization. In: 2022 International Conference on Intelligent Systems and Computer Vision (ISCV), pp. 1–8 (2022)
12. Stiennon, N., et al.: Learning to summarize with human feedback. In: Advances in Neural Information Processing Systems, vol. 33 (2020)
13. Goutom, P.J., Baruah, N., Sonowal, P.: An abstractive text summarization using deep learning in Assamese. Int. J. Inf. Technol. **15**, 2365–2372 (2023)
14. Yadav, M., Katarya, R.: A systematic survey of automatic text summarization using deep learning techniques. In: Agrawal, R., Kishore Singh, C., Goyal, A., Singh, D.K. (eds.) Modern Electronics Devices and Communication Systems. LNEE, vol. 948. Springer, Singapore (2023). https://doi.org/10.1007/978-981-19-6383-4_31
15. Helwan, A., Azar, D., Ozsahin, D.U.: Medical reports summarization using text-to-text transformer. In: 2023 Advances in Science and Engineering Technology International Conferences (ASET), Dubai, United Arab Emirates (2023)

An Improved Detection of Fetal Heart Disease Using Multilayer Perceptron

G. Someshwaran and V. Sarada$^{(\boxtimes)}$

Department of Electronics and Communication Engineering, College of Engineering and Technology, SRM Institute of Science and Technology, Kattankulathur, Chengalpattu 603203, Tamil Nadu, India
`{sg1922,saradav}@srmist.edu.in`

Abstract. Fetal Heart Disease (FHD) is a major health issue and challenge faced by the entire world in modern medicine. About 25% of babies are affected by heart disorders and 4 out of 10 have died in 2022. The desperation of FHD has become a crucial factor in the increasing rate of mortality and can even result in vulnerable consequences if not predicted at the initial stage. Most medical specialists and practising doctors have found that it's hard to anticipate and identify the disease with the existing traditional techniques at an early stage. This is because of insufficient volume of data, which has become the key failure for predicting this fetal illness and has turned out to be the most challenging task for the medical community. Since the generated data from the human body are continuous and huge in amount, the data mining techniques are utilized for an efficient classification process as it needs an accurate execution of the obtained health data. This paper attempts to reduce the risk of FHD through effective feature selection and classification-based prediction system in Ultrasound (US) images with high-performance measures and accuracy. In this implementation, the input will be obtained from the medical dataset and performs pre-processing followed by the propounded feature selection technique that efficiently decides the selection of features. Based on selected features, a novel classification is performed via Artificial Neural Network (ANN) in the detection of FHD at the early phase. Finally, the efficiency of the system is evaluated using MATLAB where the suggested system possesses an accuracy rate of 98.08% in FHD detection as normal or abnormal in an effective way.

Keywords: Fetal Heart Disease · Ultrasound · Data Pre-processing · Feature Selection · Artificial Neural Network

1 Introduction

Fetal Heart Disease (FHD) has been the substantial reason for the increasing rate of massive death worldwide for the past few decades [1] and has turned into a life-threatening disease in the medical domain which demands an accurate, reliable, and feasible system to diagnose heart disease in time.

© The Author(s), under exclusive license to Springer Nature Switzerland AG 2024
S. Satheeskumaran et al. (Eds.): ICICSD 2023, CCIS 2122, pp. 186–199, 2024.
https://doi.org/10.1007/978-3-031-61298-5_15

Fetal cardiac malformations are detected using the foremost Ultrasound (US) heart examination during the second trimester (18 to 22 weeks) of pregnancy [2, 3]. The standard fetal heart examination covers the Four-Chamber Scan (4CS) image such as the Left Atrium (LA), Left Ventricle (LV), Right Atrium (RA), Right Ventricle (RV) and Aorta as mentioned in Fig. 1, are viewed when the ultrasound beam is directed transversely to the embryo's heart.

In current, Deep Learning algorithms and techniques are indispensable in every medical sector [4–6] and can be applied to several different medical datasets to digitize and streamline huge and complex data.

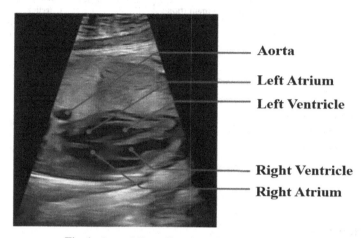

Fig. 1. Lateral Four-Chamber View of Fetal Heart

While assessing the DL technique, three distinct stages are constant: (1) Pre-Processing, (2) Segmentation, and (3) Classification as depicted in Fig. 2. Initially, the dataset is fed as input data. Then, the dataset is pre-processed where the raw data is transformed into an explicit data format. Eventually, after data pre-processing, feature selection- a segmentation approach is employed. Several feature selection techniques have been investigated by the DL algorithms [7, 8]. Amongst, the research suggests an unsupervised learning technique for feature selection, which helps find the smallest subset feature that uncovers the cluster from the medical data according to the selected criteria. An unsupervised feature selection eliminates the targeted and redundant variable via correlation. In similar, classification is equally crucial in disease prediction because an appropriate classification technique enhances the efficiency of the proposed methodology. This classification research employs the Multi-Layer Perceptron (MLP)- a feed-forward class of ANN. MLP enables prediction based on probability and classifies the items into multiple labels [9, 10]. Furthermore, MLP has the ability to learn non-linear models and train to ascertain the learned models in real-time estimation.

To summarize, our contributions are as follows:

- This article aims to reduce the probability of FHD through the use of a high-performance precise choice feature selection and a classification-based prediction technique in Ultrasound (US) images.
- Using an Artificial Neural Network (ANN), a new classification is made based on certain traits which help find FHD in its early stages.

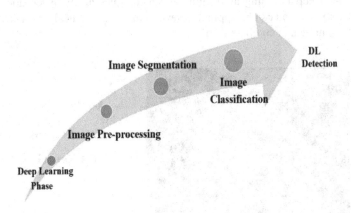

Fig. 2. Deep Learning Phase in Image Detection

2 Related Work

World Health Organization (WHO) research states, the leading cause of mortality is due to health-related disorders. Mental stress, workload and several other factors are to blame for this health problem. Accordingly, cardiac patients are treated based on the result of lab tests, the patient's medical records and the response provided by the patients. To anticipate FHD, hybrid techniques combining data mining and machine learning methods are applied [11, 12]. By utilizing these hybrid techniques, medical technicians can obtain valuable data regarding patients with FHD, enabling them to provide more accurate diagnoses. In healthcare organizations, several investigations have been done to prognose heart disorders using distinct Machine Learning (ML) and data mining approaches which are reviewed in the following sections.

Artificial Intelligence (AI) [13] has been proposed as a means of assisting in the early diagnosis of Congenital Heart Disorder (CHD) with the enumeration of Neural Networks (NN). The NN can predict congenital issues with input classes by metamorphosing inputs to pre-process integrated with segmented features and eventually, classifies the diseased elements in the ultrasound fetal image. Before being analyzed, the neural network was trained via a multi-layer net. As the outcome of the testing, the trained AI can discriminate between normal and abnormal fetal hearts.

The study [14] estimates an innovative strategy that uses a Machine Learning (ML) technique- Hybrid Random Forest with Linear Model (HRFLM) to identify specific features with the goal of increasing the precision of cardiac catastrophic detection. The detection model is laid out with distinct integration of features and numerous well-known classification methods. The result generated from the HRFLM model is 88.7% accuracy. The future enhancement of this study may be carried out using diverse amalgams of ML algorithms. In addition, new feature-selection algorithms can be created to get a wider perception of the relevant traits as an upgrade to improve heart disease prediction performance.

Reddy et al. [15] generated a primary model called Adaptive Genetic Algorithm via Fuzzy Logic (AGAFL) to detect cardiac disorder at an initial stage. To begin, the affected feature is extracted using the rough set theory and subsequently, the AGAFL classifier is utilized in detecting heart disease from the selected feature. Further, the prediction result is proved to be efficient when compared with the other dataset findings from existing methods that are formerly accessible to the public. This algorithm can be improved by implementing a meta-heuristic method to acquire an optimal outcome.

Babu et al. [16] laid out the hybrid methods for detecting different sorts of heart disease by combining the Grey Wolf Optimization with an auto-encoder based on Recurrent Neural Network (GWO + RNN). The GWO is utilized to extract the features and the detection of disease is done by RNN which has resulted in better performance. Several datasets, including Cleveland, Mammographic, and Hungarian, are employed, and their findings were compared with the proposed model. The outcome obtained from the model outlays an increase in accuracy by 16.82% than other existing methods. Improvement can be done by implementing various techniques to attain more accuracy than the current procedures.

The grey wolf algorithm has also been applied for predicting fine results from various supplemental medical datasets. Similarly, S. S. G et al. [17] propounded a prognostic model for detecting diabetes via the fuzzy rule concept and GWO method. The model's base has been developed and worked with the ant colony optimization technique with an accuracy of 71%. The level of precision must be improved in future diagnoses.

Eventually, Salem et al. [18] employed the Genetic Algorithm (GA) to investigate various models for detecting heart disorders and for selecting features of the fetal heart. The GA is used in an optimized way to get better evaluation results than other traditional methods. The performances were retested with different datasets of heart disease and evaluated in real-time with four distinct classifiers namely Random Forest (RF), Decision Tree (DT), Navies Bayes (NB) and Support Vector Machine (SVM) where Navies Bayes has resulted to be preferable than the other classifiers with respect to the data set and experimental outcomes.

The key focus [19] is to significantly optimize the feature selection by integrating SVM with Genetic Algorithm (GA). The model findings disclosed an accuracy rate of 88.34% with the selected features in heart disease prediction when compared with existing feature selection concepts such as Correlation-based Feature Selection (CFS), Chi-Squared, Filtered Subset, Info Gain and Consistency Subset. Furthermore, the technique has proved that ROC analysis provides efficient performance in the SVM classifier where the model's efficiency is implemented in MATLAB software with the dataset collection in Cleveland Heart Disease (CHD) data image.

Patro et al. [20] employed a diverse mix of machine learning methods to improve the classification and feature selection for better prediction of heart disease. Utilizing the optimization technique, the optimal data from the existing dataset have been determined. Subsequently, the classification algorithm screens the data with similar classes using mathematical tools with computerized techniques and enables the identification model to be analyzed for training and testing the data. The framework of the proposed model has been designed by considering the outcomes generated by different classifiers such as Bayesian Optimized Support Vector Machine (BO-SVM), Navies Bayes (NB), Salp Swarm Optimized Neural Network (SSA-NN) and K-Nearest Neighbors (K-NN). Based on the UCI dataset, BO-SVM performs better than the other traditional techniques with an accuracy of 93.3% and a sensitivity of 80%. The future scope concerns the deep learning algorithm can be replaced as an enhancement for high efficiency.

Shalini et al. [21], encountered that numerous authors have introduced diverse Machine Learning (ML) methodologies that primarily focused on disease detection and safeguarding fetal life via the medical professions. Despite the use of several ML techniques, certain flaws exist due to inadequate proportions of the heart and body. The deployment of deep learning methods today, especially in the health sector, has greatly aided medical practitioners in their efforts to safeguard human lives. The suggested DL method has reviewed the prediction of heart diseases in a hybrid way using Sequential Minimal Optimization (SMO) in enumeration with the ANN method and has outperformed other traditional techniques by enhancing the efficiency rate of 92.45% accuracy with an improvised classification metrics in disease detection. The author suggests that accuracy may be improved by using a large number of datasets and incorporating various feature selection strategies.

Hence from the above-related work, ANN classifier is an appropriate scheme when a given input attribute is more. Compared to many other classification techniques this algorithm is simple and has better performance. Patients with FHD characteristics can be easily identified by the ANN technique more appropriately.

3   Dataset

3.1   CHD Dataset

To be fed into the training framework, data must be processed initially to extract the necessary features from the appropriate dataset. For this study, the dataset of Cleveland Heart Disease (CHD) [22] has been considered and it is segregated into two sections.

1. Build and train the learning model.
2. Testing in order to ascertain the model's accuracy.

The CHD dataset is flawless and has no omissions. Hence, no other processing steps are needed besides formalization. Based on the datasets of CHD, the original data is collected, coded, and then assigned to the classification.

4   Methodology and Models

In most recent surveys, numerous congenital heart disorder prediction techniques are presented, however, have certain limitations. Hence, developing an accurate model for identifying FHD becomes crucial. Consequently, FHD prediction has been proposed effectively by employing the selection of features and classification which necessitates an optimized unsupervised method for selecting the relevant traits and an innovative ANN for classifying the selected feature for precise disease detection. The methodology comprises three distinct phases, which are,

1. Data pre-processing,
2. Feature selection and
3. ANN Classification.

The dataset of Fetal Heart Disease is acquired as input data for the suggested methodology. Prior to using the feature selection technique, the provided data set is pre-processed. An algorithm known as unsupervised selection of features is used in determining the feature in an optimum mode. The proposed Multi-Layer Perceptron (MLP) Algorithm has been used as a classification method for identifying FHD after choosing the appropriate feature. The flow chart of the developed model is portrayed in Fig. 3.

By utilizing ANN with feature extraction, the proposed approach avoids the need for manual and complex feature capabilities which can be challenging and time-consuming for medical image analysis tasks. However, ANN's Multi-Layer Perceptron obtains an adequate and diverse dataset in an optimal way for the successful training and generalization of the model than other traditional techniques.

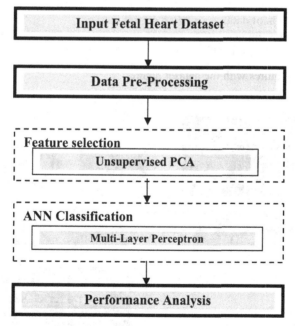

Fig. 3. Overall Proposed Approach

5 Data Pre-processing

Pre-processing of data entails formatting the raw data and unprocessed information into a readable data format and pertains to the data removal or manipulation prior to their use to enhance or ensure performance. It is also a key aspect in the case of data mining methods, where the raw information of data seems to be difficult. Since real-time image data are unclear and complex, data pre-processing helps reduce the data complexity while analyzing the data and it checks the data quality before applying DL or data mining algorithms. The efficacy and performance metrics of the FHD detections are not only concluded by the utilization of the DL algorithmic calculation but also focus on the dataset integral quality via pre-processing methodology. The obtained datasets may have missing data, errors, noise, redundancy, dataset size and quite a few issues which may render the data unsuitable for direct use with the algorithm of deep learning. Moreover, several datasets with numerous factors make the relevant data difficult to analyze the algorithm, pattern discover and accurate predictions. The issues can be eliminated by data analysis and by applying a suitable data pre-processing model.

Figure 4 illustrates the pre-processing phase which signifies the generated ultrasound datasets to image enhancement state i.e., provides in an affordable form that the algorithm understands. The steps involved in data pre-processing techniques are:

1. Noise removal and data enhancement

    - Cleaning of data,
    - Transformation of data,
    - Missing values imputation,
    - Normalization of data.

2. Extraction of features with the dataset nature.

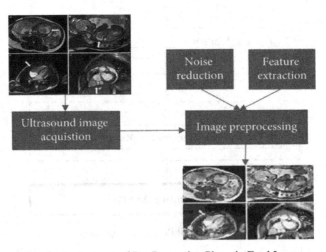

Fig. 4. An Instance of Pre-Processing Phase in Fetal Image

The proposed pre-processing model gathers the dataset of fetal heart ultrasound images along with corresponding diagnostic labels i.e., normal or abnormal. It pre-processes the images, including resizing to a consistent size, normalization i.e., scaling pixel values to [0, 1], and augmentation to increase the dataset size i.e., rotation, flipping, and zooming for better generalization. Eventually, splits the preprocessed data into training and testing sets.

5.1 Feature Selection

The acquired dataset incorporates an immense volume of data concerning multiple features and selecting only the best of these characteristics will be assessed by the DL model which computes the variations between different adequate models that intend with refined measures, intense computational efficiency and substantial accuracy. The feature selection is widely categorized into supervised, unsupervised and semi-supervised methods. In supervised method, the information of the provided classes will be included in the selection but the unsupervised method will use the task with ground truth labelling and select the discriminative feature effectively for differentiating the samples from other classes. The research proposes an unsupervised Principle Component Analysis (PCA) feature selection as a guideline for working towards the data to obtain improved outcomes with maximum relevance and minimum redundancy. This feature selection method is exploited after data pre-processing for obtaining the reduction of data and helps to attain the data model in an efficient way which typically results in improved efficacy, increased precision in learning, low computation costs and better interpretation of the model. A firmly selected and feasible feature PCA selection always optimizes the subset of the feature and uproots the original conditional distribution of the target. Moreover, it designates the most pertinent, consistent and non-redundant traits. PCA function in feature selection for fetal heart disease diagnosis highlights significantly deeper trends in the data while assisting in dimensionality reduction. However, it develops a useful and precise labelling model for fetal cardiac disease diagnosis in conjunction with other approaches, such as domain knowledge, medical expertise, and task-specific feature selection methods. The objectives of the PCA feature selection methodology involve:

1. Model simplification to make easier interpretation by the users,
2. Shorter duration for training,
3. Neglecting the dimensionality flaw,
4. Improving the generalization by decreasing overfitting. i.e., Variance reduction.

5.2 Image Classification

Classification in Deep learning distinguishes the data points into several classes, allowing the selected features from datasets of all sorts involving simple, large and intricate datasets. The classification notably utilized the ANN's Multi-Layer Perceptron (MLP), which can be promptly modified to enhance data intensity. MLP classification's primary objective is to link the interested variable with the needed variable, i.e., provides the connection between the prediction and the variable, where the variable of interest is based on a qualitative analysis in FHD detection. The MLP algorithm employed in the

DL technique is defined as classifiers whereas the result obtained by this algorithm is called instances. MLP is the supplement of the feed-forward network with three stages:

1. The primary layer is the input layer,
2. The intermediate layer is labelled as the hidden layer and
3. Finally, the output layer.

Figure 5 illustrates the complete structure of the MLP network and describes the different phases of MLP where the selected feature is fed as the input image. The input layer accepts the appropriate features in the form of arrays and applies a matrix filter which twists and alters the data to detect the diseased pattern of the fetal heart image. The hidden layer employs the convolutional operation to get the rectified map of the diseased image. Finally, the output layer provides the experimental outcome of FHD detection as normal or abnormal in an optimal way. MLP neural network utilizes an error back-propagation algorithm for learning its variable and parameter. This technique comprises two different phases via layers of neural network named, the backward pass and forward pass. In the forward phase, the network is applied with the input pattern for obtaining the output set with fixed weights. In the backward phase, the connection weights will be propagated back by errors and the weights will get updated to minimize the prediction error. MLP has the ability to learn new models in online and has the capability to learn non-linear models. The output of the MLP is obtained based on inputs, weights and biases.

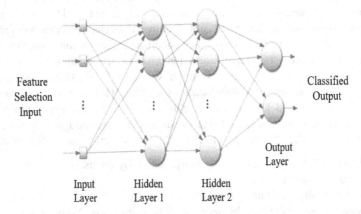

Fig. 5. Multi-Layer Perceptron Architecture

The Multilayer Perceptron (MLP) classifier is an effective method for inputs with more attributes. The approach is simpler and performs better than other ML/DL classification methods. The MLP classifier algorithm more accurately identifies patients with FHD traits.

6 Fetal Heart Anatomical Findings

Gray-Level Co-Occurrence Matrix (GLCM) measures aid in FHD test and prognosis. In the context of fetal heart disease diagnosis from medical images, GLCM is applied to extract texture features from ultrasound images of the fetal heart. These features helps to characterize different tissue structures or anomalies, which is useful in building diagnostic models. Subsequently, the backpropagation algorithm trains the neural network based on the dataset of features extracted from medical images, including texture features obtained from GLCM. The neural network is then used as a classifier to predict the presence or absence of fetal heart disease in ultrasound images. Consequently, trained and modified MLP is applied to find fetal heart diseases in Apical 4 chamber (A4C) screening images and used selected features to calculate the Cardiac Axis (CA) and Fetal Abdominal Circumference (FAC) labelled for each chamber of the fetal heart. The research represents Normal, Ventricular Septal Defect (VSD), Tetralogy of Fallot (TOF), and Hypoplastic Left Heart Syndrome (HLHS) hearts from CHD to be estimated in the training and testing phase are illustrated in Fig. 6.

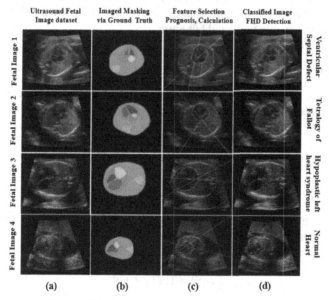

Fig. 6. The anatomical image of the fetal heart masking, feature selection and ANN classification. 6(a): Fetal heart ultrasound input images of diverse datasets, 6(b): Ground truth labelling of fetal anatomical structures: Colour red-left ventricle, colour pink-left atrium, colour blue-right ventricle, pale blue-right atrium, 6(c): Feature Selection prognosis calculation, 6(d): Image classification with disease prediction.

7 Performance Metrics Estimation

Table 1 depicts the enactment of the FHD prediction method based on the examined data of the stated systems.

The projected method results on the source of several testing data where the parameters are evaluated and compared on the basis of the performance metrics such as

- Accuracy,
- Precision,
- Recall and
- F1-Score.

7.1 Accuracy

Accuracy determines the prophecy fractions achieved precisely via the selected features which resolve the prognosis of heart disease in an explicit manner and is illustrated by Eq. (1).

$$Accuracy = TP + TN / TP + TN + FP + FN \tag{1}$$

Where TP signifies True Positive which is accurately predicted and abnormal. Next, TN is labelled as True Negative, which is appropriately predicted and normal. The term FN denotes False Negative where predictions are incorrect and abnormal. FP refers to False Positive that is incorrectly predicted but normal.

7.2 Precision

Precision depicts the proportion of tuples through which the precise prediction is done for the heart disease of abnormal patients and is premeditated by Eq. (2) that follows.

$$Precision = TP / TP + FP \tag{2}$$

7.3 Recall

Recall deduces the fraction of tuples, where the predictions are made accurately for abnormal patients by recognizing the precise individuals with heart-related diseases. Recall calculation is determined in Eq. (3).

$$Recall = TP/TP + FN \tag{3}$$

7.4 F1-Score

F1-score is determined by the harmonic mean of precision and recall. F1-score precisely classifies numerous feasible positive samples, rather than maximizing the correct classification numeral. The computational Eq. (4) is as follows Fig.7.

$$F1 - Score = (2 * precision * recall) / (precision + recall) \tag{4}$$

Table 1. Performance Analysis

ML/DL Methods	Testing data	TP	TN	FP	FN	Precision	F1-Score	Recall	Accuracy %
HRFLM	122	52	8	10	52	0.85	0.85	0.85	88.7
SVM	104	49	2	10	43	0.89	0.88	0.88	88.34
BO-SVM	53	30	0	4	19	0.93	0.92	0.92	93.3
SMO	82	40	1	8	33	0.9	0.89	0.89	92.45
Proposed MLP	110	104	1	0	5	0.96	0.96	0.96	98.08

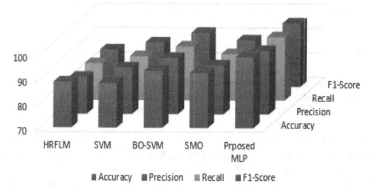

Fig. 7. Comparison Graph of Distinct ML/DL Algorithms in accordance to Accuracy, Precision, Recall and F1-Score

8 Results Analysis

Table 1 estimates the parametric values of FHD with distinct prediction systems and the proposed MLP model via testing data in reference to accuracy, precision, recall and f1-score where the HRFLM model results in the accumulation of 122 testing data with 88.7% accuracy, 0.85 for precision, recall and f1-score. The second SVM-an ML algorithm sequels with 104 testing data and acquired 88.34% accuracy, 0.89 precision, 0.88 recall and f1-score. Subsequently, BO-SVM and SMO accumulate with the testing data of 53 and 82 which results in comparable accuracy of 93.3% and 92.45% and ensuing with the precision, recall and fi-score statistics of 0.93 for BO-SVM and 0.89 for SVM in FHD prognosis (Fig.7).

The tabulation result depicts that the proposed MLP is the finest model for congenital heart disorder detection which is acquired with the testing data of 110 in enumeration to TP and FP values of 104 and 0, TN and FN values of 1 and 5 with an estimated accuracy of 98.08% and precision, recall and f1-score of 0.96. Hence, the research has exposed that the planned MLP method has more competence than the other existing methods in the FHD prediction with the stated parameters.

9   Conclusion

FHD is the leading cause of mortality rates, and it should be diagnosed appropriately at an early stage for proper treatment. The algorithms of DL and Meta-heuristic techniques are one among the tools used for disease diagnosis. In such a way, the proposed methodology utilized DL in the detection of FHD at a premature phase where the input data from the Cleveland dataset is pre-processed and fed to optimized PCA feature selection that designates the appropriate feature and resolved the issues by extracting the information and groups the data with ground truth labels. Finally, the ANN classifier detected FHD with parametrical measures of 98.08% accuracy. In order to ascertain and validate the efficacy of the classifier, the suggested MLP has been compared with distinct ML and DL algorithms, such as the HRFLM, SVM, BO-SVM and SMO algorithms. The performance evaluation of the suggested method is analyzed using MATLAB where the outcomes proved with an improved efficiency of the proffered MLP algorithm.

References

1. Bernier, P.-L., Stefanescu, A., Samoukovic, G., Tchervenkov, C.I.: The challenge of congenital heart disease worldwide: epidemiologic and demographic facts. Semin. Thorac. Cardiovasc. Surg.: Pediatric Card. Surg. Ann. **13**(1), 26–34 (2010). https://doi.org/10.1053/j.pcsu.2010.02.005
2. Edupuganti, M., Rathikarani, V., Chaduvula, K.: A real and accurate ultrasound fetal imaging based heart disease detection using deep learning technology. Int. J. Integr. Eng. **14**(7) (2022). https://doi.org/10.30880/ijie.2022.14.07.005
3. Sapitri, A.I., Nurmaini, S., Rini, D.P., Rachmatullah, M.N., Darmawahyuni. A., Gusendi, A.: Detection of fetal cardiac chamber three vessel trachea view using deep learning. In: 2022 9th International Conference on Electrical Engineering, Computer Science and Informatics (EECSI), Jakarta, Indonesia, pp. 43–48 (2022). https://doi.org/10.23919/EECSI56542.2022.9946528
4. Liang, S., Li, Q.: Automatic evaluation of fetal heart rate based on deep learning. In: 2021 2nd Information Communication Technologies Conference (ICTC), Nanjing, China, pp. 235–240 (2021).https://doi.org/10.1109/ICTC51749.2021.9441583
5. Rajamhoana, S.P., Devi, C.A., Umamaheswari, K., Kiruba, R., Karunya, K., Deepika, R.: Analysis of neural networks based heart disease prediction system. In: 2018 11th International Conference on Human System Interaction (HSI), Gdansk, Poland, pp. 233–239 (2018). https://doi.org/10.1109/HSI.2018.8431153
6. Shankar, H., et al.: Leveraging clinically relevant biometric constraints to supervise a deep learning model for the accurate caliper placement to obtain sonographic measurements of the fetal brain. In: 2022 IEEE 19th International Symposium on Biomedical Imaging (ISBI), Kolkata, India, pp. 1–5 (2022). https://doi.org/10.1109/ISBI52829.2022.9761493
7. Venkatesh, B., Anuradha, J.: A review of feature selection and its methods. Cybern. Inform.Technol. **19**(1), 3–26 (2019). https://doi.org/10.2478/cait-2019-0001
8. Sutha, K., Tamilselvi, J.J.: A review of feature selection algorithms for data mining techniques. Int. J. Comput. Sci. Eng. **7**(6), 63 (2015)
9. Deepika, D., Balaji, N.: Effective heart disease prediction using novel MLP-EBMDA approach. Biomed. Signal Proc. Control **72**, 103318 (2022). https://doi.org/10.1016/j.bspc.2021.103318

10. Bhoyar, S., Wagholikar, N., Bakshi, K., Chaudhari, S.: Real-time heart disease prediction system using multilayer perceptron. In: 2021 2nd International Conference for Emerging Technology (INCET), Belagavi, India, pp. 1–4 (2021). https://doi.org/10.1109/INCET51464.2021.9456389
11. Francis, F., Wu, H., Luz, S., Townsend, R., Stock, S.: Detecting intrapartum fetal hypoxia from cardiotocography using machine learning. Comput. Cardiol. (CinC) **49**, 1–4 (2022). https://doi.org/10.22489/CinC.2022.339
12. Mehbodniya, A., Lazar, A.J.P., Webber, J., et al.: Fetal health classification from cardiotocographic data using machine learning. Expert Syst. **39**(6) (2021). https://doi.org/10.1111/exsy.12899
13. Alan, M., Aküner, M.C., Kepez, A.: Biosignal classification and disease prediction with deep learning. In: 2020 Innovations in Intelligent Systems and Applications Conference (ASYU), Istanbul, Turkey, pp. 1–5 (2020).https://doi.org/10.1109/ASYU50717.2020.9259852
14. Mohan, S., Thirumalai, C., Srivastava, G.: Effective heart disease prediction using hybrid machine learning techniques. IEEE Access **7**, 81542–81554 (2019). https://doi.org/10.1109/ACCESS.2019.2923707
15. Reddy, G.T., Reddy, M.P.K., Lakshmanna, K., Rajput, D.S., Kaluri, R., Srivastava, G.: Hybrid genetic algorithm and a fuzzy logic classifier for heart disease diagnosis. Evol. Intel. **13**(2), 185–196 (2019). https://doi.org/10.1007/s12065-019-00327-1
16. Babu, S.B., Suneetha, A., Babu, G.C., Kumar, Y.J.N., Karuna, G.: Medical disease prediction using Grey Wolf optimization and auto encoder based recurrent neural network. Periodicals Eng. Nat. Sci. (PEN) **6**(1), 229 (2018). https://doi.org/10.21533/pen.v6i1.286
17. Siva Shankar, G., Manikandan, K.: Diagnosis of diabetes diseases using optimized fuzzy rule set by grey wolf optimization. Pattern Recognit. Lett. **125**, 432–438 (2019). https://doi.org/10.1016/j.patrec.2019.06.005
18. Salem, T.: Study and analysis of prediction model for heart disease: an optimization approach using genetic algorithm. Int.J. Pure Appl.Math. **119**, 5323–5336 (2018)
19. Gokulnath, C.B., Shantharajah, S.P.: An optimized feature selection based on genetic approach and support vector machine for heart disease. Clust. Comput. **22**(S6), 14777–14787 (2018). https://doi.org/10.1007/s10586-018-2416-4
20. Patro, S.P., Nayak, G.S., Padhy, N.: Heart disease prediction by using novel optimization algorithm: a supervised learning prospective. Inform. Med. Unlocked **26**, 100696 (2021). https://doi.org/10.1016/j.imu.2021.100696
21. Shalini, P.K.S., Sharma, Y.M.: An intelligent hybrid model for forecasting of heart and diabetes diseases with SMO and ANN. In: Intelligent Energy Management Technologies: ICAEM 2019, pp. 133–138. Springer, Singapore (2020). https://doi.org/10.1007/978-981-15-8820-4_13
22. https://www.kaggle.com/datasets/cherngs/heart-disease-cleveland-uci

Application of Deep Learning Techniques for Coronary Artery Disease Detection and Prediction: A Systematic Review

M. Jayasree$^{(\boxtimes)}$ and L. Koteswara Rao

Department of Electronics and Communication Engineering, Koneru Lakshmaiah Education Foundation, Hyderabad, India
{jayasree.kaluvoju,koteswararao}@klh.edu.in

Abstract. Coronary artery disease, a leading cause of death, occurs when coronary arteries narrow due to plaque or blood clots. It is possible to visualize these blood clots in the blood vessels through angiograms, a medical imaging technique. An angiogram is a type of medical imaging procedure used to see where blood clots or plaques have formed inside the coronary arteries. The conventional coronary angiogram or X-ray angiogram has been measured as a gold standard to predict coronary artery disease. Machine learning and deep learning techniques are widely deployed in many applications that include feature extraction, enhancement, and pattern analysis. The development of technology and the accessibility of vast amounts of data have made it relatively simple to analyze the condition of the coronary arteries, identify the disease, and make an early diagnosis. In this review paper, coronary artery disease assessment, angiograms, and various deep learning methods for coronary artery disease prediction are reviewed, and the assessment results are showcased.

Keywords: Coronary Artery Disease · Angiogram · Stenosis · Convolutional Neural Networks · Deep Learning

1 Introduction

Heart attacks, a common cause of death, are often referred to as coronary artery disease (CAD), also known as cardiovascular disease. The human heart, the strongest muscle, beats approximately 100,000 times per day over a 70-year lifespan. The uterine heart frequently begins to beat between 21 and 28 days after conception, long before the baby is born. Heart disease is a significant risk factor for postmenopausal women, males over 40, and younger generations with diabetes. Factors like obesity, high cholesterol, smoking, and a family history of heart attack and high blood pressure can cause plaque to form in coronary arteries, blocking blood flow [1, 2]. Congenital heart disease, a heart abnormality from birth, is the most common cause of unexpected cardiac arrests. As technology grows day by day, human beings are also changing their food habits, working style, stress at work, restlessness, and other elements that add to heftiness [3, 4].

© The Author(s), under exclusive license to Springer Nature Switzerland AG 2024
S. Satheeskumaran et al. (Eds.): ICICSD 2023, CCIS 2122, pp. 200–211, 2024.
https://doi.org/10.1007/978-3-031-61298-5_16

The coronary artery supplies oxygen-rich blood and nutrients to the heart. Blockages can restrict blood flow, requiring treatment with a stent or bypass surgery. Early diagnosis and treatment can reduce disease severity, but understanding risk factors and improving diagnosis accuracy is crucial. If any plaque or blockage is formed, blood flow will be restricted in the coronary arteries as shown in Fig. 1. To treat this narrowed artery a stent (tiny mesh tube) can be placed in the narrowed artery which helps to enlarge the narrowed artery and improves the blood flow. If a person is having more blockages in the arteries or one of the larger arteries is blocked then bypass surgery can be done. The severity of a disease can be reduced through early diagnosis and treatment. To improve the accuracy of diagnosis, it is essential to get a clear understanding of risk and preventative factors [5, 6].

CAD diagnosis involves invasive or non-invasive methods, such as ECG, MRI, echocardiography, stress testing, and SPECT, depending on disease severity, impact on cardiac function, and treatment options. Invasive or non-invasive techniques are used to assess the disease's severity, how it affects cardiac function, and the best course of treatment for the patient. Angiogram is a costly and complex invasive method used to diagnose heart disease, detecting blockages in arteries in various parts of the body. It is not suitable for large population screening or close treatment monitoring. Angiogram is a procedure used to identify blockages in arteries in various parts of the body, including the heart, using X-ray imaging to detect and monitor vascular anomalies.

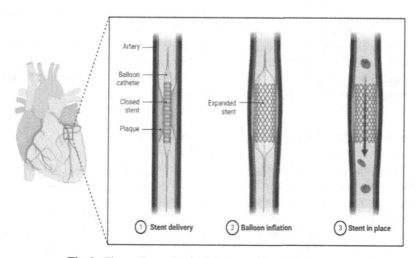

Fig. 1. Plaque Formation in Arteries and Stent Replacement

The computational power and data storage capabilities of intelligent computing systems are attracting the researchers to perform in-depth analysis of biomedical data. The growth of artificial intelligence (AI), particularly machine learning algorithms, in diagnostics, risk assessment, and preventive measures is the result of the convergence of computer power and data storage capacity. Machine Learning (ML)-based AI has been successfully applied in fields like dermatology and ophthalmology, and now it is gradually making its way to the field of cardiology.

Recent machine learning advancements, particularly deep learning, have significantly improved the recognition, categorization, and quantification of patterns in medical images. Deep learning is gaining popularity in medical applications due to its superior feature extraction capability, which is learned exclusively from data rather than manually created features. Deep learning is rapidly gaining traction as the leading foundation in various medical applications due to its improved outcomes.

Deep learning's success is attributed to the development of learning algorithms, the availability of large amounts of big data, and advancements in high-tech GPUs. Deep learning is considered an improvement over traditional artificial neural networks due to its ability to construct more number of layers. As a result, deep learning has garnered a lot of attention among researchers in the study of medical images. Compared to a simple neural network, a deep neural network approximates the human brain much more closely using sophisticated mechanisms.

A few models were developed to detect stenosis, plaque in the coronary arteries, and artery segmentation with different algorithms. R. Gharleghi [30] introduced the CTCA (Computed Tomography Coronary Angiography) method to estimate coronary artery anatomy and disease accurately but failed to detect the disease with time efficiently. Jong Hak Moona [31] developed a convolutional neural network for automatic stenosis identification using coronary angiography images and classification was improved but unable to provide diagnostic inspection for medical professionals. It ia also noted that accuracy was compromised and for multiresolution images. Deep Learning Image Reconstruction (DLIR) was proposed by Domenico De Santis [32] with the aim of assessing correlation to improve accuracy and performance, but it doesn't perform well in the presence of noise. The organization of this paper is as follows: Sect. 1 explains the introduction of the paper; Sect. 2 talks about Deep Learning Architectures and Applications; Sect. 3 provides performance metrics; Sect. 4 focuses on Deep Learning for CAD Prediction; and Sect. 5 is the conclusion of the paper.

2 Deep Learning Architectures and Applications

Deep Learning (DL) is increasingly popular in artificial intelligence (AI) due to its superior predictions, making machines more intelligent devices. Deep learning architecture employs a significant amount of data to train an artificial neural network. Deep learning networks require extensive data for self-learning and accurate predictions, while conventional machine learning (ML) consists of algorithms for data splitting and prediction. The researchers utilized various datasets, including Microsoft Common Objects in Context (COCO), ImageNet, CIFAR-10, MNIST, and FERET, to make intelligent decisions. Deep learning architectures consist of three layers: input, hidden, and output, with more hidden layers in deep neural networks. Each layer in a convolutional neural network (CNN) is fully connected, with input, pooling, and activation layers. These neural networks are used to solve challenging issues. Figure 2 [18] depicts the CNN architecture. Autoencoders include feed-forward neural networks. The data is then decoded from the hidden layers after the input has been encoded. Auto-encoders are used for document analysis, natural language processing and face recognition. Recurrent neural networks (RNN) are used to implement Long Short-Term Memory (LSTM). The

Restricted Boltzmann Machine (RBM), an unsupervised machine learning algorithm [14], the RBM architecture has visible and hidden layers and learns the probability distribution of applied input.

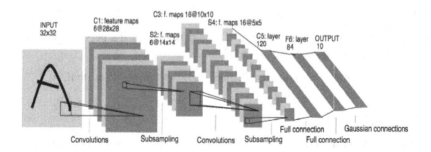

Fig. 2. LeNet CNN Architecture [18]

Deep learning techniques have made substantial progresses in biomedical applications by providing more accurate results in the detection of number of diseases. Few notable applications are cancer detection, diabetic retinopathy, Alzheimer's disease. It is also widely deployed in biomedical image denoising, segmentation and reconstruction. In the risk assessment and prediction, deep learning techniques are used for critical clinical decisions and cardiovascular disease detection and prediction. Deep learning has shown significant promise in various aspects of heart disease detection, diagnosis, risk assessment, and treatment. Heart disease detection and prediction are used in the assessment of coronary artery disease, heart failure, and cardiac arrhythmias, heart disease-related techniques requires large and diverse datasets and heavy computational elements to ensure patient safety and clinical effectiveness. In image analysis based deep learning models, CAD detection can be performed by analyzing coronary angiography, CT angiography (CTA), and cardiac MRI images.

3 Performance Metrics

Healthcare decisions matter greatly because we need to understand precisely when a model applies to a patient and when it does not. For an algorithm to perform well, particularly in the fields of deep learning and machine learning, the following crucial elements are required:

3.1 Accuracy

The metric that analyzes a model's efficiency and determines whether it is correctly classified or not classified is called accuracy.

$$Accuracy = \frac{Examples\, correctly\, classified}{Total\, number\, of\, examples}$$

$$Accuracy = P(correct)$$

3.2 Sensitivity or P (+/Disease)

It represents the chances that a model will predict positively if a patient has a diagnosis, also known as the true positive rates.

3.3 Specificity or P (-/Disease)

It represents the chances that a model will predict negatively if a person doesn't have a disorder, or the true negative rate of a model.

$$Accuracy = Sensitivity*P(disease) + Specificity*P(normal)$$

3.4 Prevalence

Prevalence is the probability that a patient will have a disease in the population.

$$Prevalence = P(disease) = \frac{populationwithdisease}{Totalnumberofpopulation}$$

$$Accuracy = Sensitivity*Prevalence + Specificity*(1 - Prevalence)$$

3.5 Positive Predict Value (PPV)

$$PPV = P\left(\frac{disease}{+}\right) = \frac{P(diseaseand+)}{P(+)}$$

3.6 Negative Predict Value (NPV)

$$PPV = P\left(\frac{normal}{-}\right) = \frac{P(normaland-)}{P(-)}$$

3.7 Confusion Matrix

The performance of a classifier can be represented using a confusion matrix, as shown in Table 1. The terminology used to describe a confusion matrix can be difficult to understand. Error matrix and matching matrix are other names for confusion matrix. The model's output is the predicted class, and the true class in this case is often referred to as the ground truth (GT) values. These values allow us to define the metrics as

$$Sensitivity = \frac{TP}{TP + FN}$$

$$Specificity = \frac{TN}{FP + TN}$$

$$PPV = \frac{TP}{TP + FP}$$

$$NPV = \frac{FN}{FN + TN}$$

Table 1. Confusion matrix

		P	N
Predicted class	P	TP	FP
	N	FN	TN

3.8 ROC Curve

Receiver Operating Characteristic (ROC) curve determines how well categorization models are functioning. It permits us to visibly plot the connection between a model's sensitivity and specificity at various decision thresholds. When the probability of a patient having a condition reaches a certain threshold, we refer to this as positive evidence. A patient is considered to have no disease if the disease probability is below a certain threshold or is interpreted negatively.

3.9 F-Score or F-Measure

This is a main indicator to decide how well the classification procedure is performed. The F-Score takes into consideration the classification procedure's recall as well as precision. The higher the F-Score value better the classification procedure. Generally, F-score lies in between 0 and 1 i.e. $0 < F < 1$

$$F = 2 * \frac{Precision * Recall}{Precision + Recall}$$

$$Precision = \frac{TP}{TP + FP}$$

$$Recall = \frac{TP}{TP + FN}$$

The portion of accurate positive classification is recall.

4 Deep Learning for CAD Prediction

In order to extract vessels from X-ray angiograms using deep learning, Nasr-Esfahani et al. [21] preprocessed the image using the top-hat transform for bright and dim regions. A convolutional neural network (CNN) with convolve and pooling layers is used to divide the image into vessel and background areas and calculate each pixel's likelihood of being in a vessel region. Simson et al. reconstructed high-quality ultrasound images from subsampled data using the "DeepFormer" method, which had a lower computing cost and higher lateral resolution. Because of QuickNat's gradient flow, the reconstruction procedure makes use of its fully convolutional network design.

In situations where traditional decision support systems are ineffective, R. Chitra et al. created an intelligent framework for predicting heart illness using cascaded neural networks (CNN). Using three-dimensional coronary computed tomography angiography (CTA) images and a fully convolutional neural network (FCNN), Ye Shen et al. improved accuracy by using attention gate and level set techniques. The 3D FCNN incorporates level set functions to enhance network prediction performance and smooth coronary artery segment boundaries through careful gating.

Koti et al. introduced the Harmony Search (HM-L) strategy, utilizing Lévy conveyance, to accurately predict cardiac diseashybrid e, reducing unnecessary features in the image. The selection of features is then applied to classification models. The author proposes the Levi distribution as a mathematical model to initiate sudden drift due to issues involved in the Harmony-Search algorithm.

Gokhan A. et al. utilized Deep Belief Networks (DBN) to introduce a novel diagnostic method for cardiologists. The new 15-s-long diagnosis system uses a moving window evaluation of short-term ECG signals to differentiate between patients with and without CAD. The Hilbert-Haung change is utilized for handling transient ECG signals due to its productivity in handling non-fixed and non-straight signals. The Hilbert-Haung transform is utilized in deep learning techniques to determine the instantaneous frequency components of a signal using IMF and DBN classifications. This model outperforms other neural network models in terms of accuracy rates.

In order to estimate the risk of coronary artery atherosclerosis in the UCI and STU-LONG datasets, Nikan et al. applied machine learning. This study emphasizes the value of early identification in lowering mortality. Clinical datasets that have missing data are presented using the Ridge Expectation Maximization Imputation (REMI) methodology, which also decreases the dimensionality of the feature space. Powerful classifiers like Extreme Learning Machines (ELM) and Support Vector Machines (SVM) are used to forecast the CAD of no-CAD labels for specific persons.

The exploitative and exploratory capabilities of GOA were used by Heidari et al. to create a hybrid stochastic training strategy for multilayer perceptron neural networks. The five main medical conditions that affect individuals with breast cancer, Parkinson's disease, diabetes, coronary artery disease, and orthopaedics are comprehensively analyzed using this model. Al-Milliet et al. [3] introduced a supervised method for creating multilayer neural networks that employs error-back propagation algorithm, which involves forward and backward passes. The method applies activity patterns to both forward and backward passes using error correction learning principles. Error-correcting and generalized delta rules are used to alter the weights of networks to ensure precision and consistency.

Zotti et al. [28] introduced a novel CNN architecture for segmenting images from short-axis CMRI slices. The system segments the left ventricle's epicardium, endocardium, center, and right ventricle using raw images without pre-processing. It uses a multi-resolution grid-net architecture to determine cardiac region boundaries. Chung developed two segmentation techniques for angiography, with the first focusing on objectively identifying and measuring vascular abnormalities. The use of augmented vessels, a computer-generated vessel, is often employed to estimate a portion of post-treatment

Table 2. Various techniques used for coronary artery disease detection

Author	Technique Used
Kim & Kang, 2017 [15]	Neural 1networks with Feature Correlation Analysis
Subhadra & Vikas, 2019 [27]	Multi-layer Perceptron Neural Network with Back-propagation
Serife Kaba [9]	Densenet, Efficient Net, ResNet, MobileNet
Abdolmanafi et al., 2021 [1]	Auto encoders
Han et al., 2020 [12]	Convolutional Neural Networks
Gessert et al., 2019 [11]	ResNet50-V2, DenseNet121
Liu et al., 2020 [20]	multi-scale convolution neural networks
Fischer et al., 2020 [10]	RNN-LSTM
Lee et al., 2021 [19]	Inception Resnet v2, VGG, and Resnet 50
Zreik et al., 2019 [29]	Recurrent CNN

vessel lumens. The second method enhances angiogram visualization and DVR quality by utilizing vessel boundary information and a Hessian-based image enhancement technique. Table 2 lists various methods for detecting coronary artery disease.

A high-performing automated approach for assessing the Agatston calcium score in general-indication non-contrast chest CT scans was created by Shadmi, Bregman-Amitai, Mazo, and Elnekave et al. The authors assess U-Net and Dense Net topologies, which have a Pearson correlation coefficient of 0.98 and predict higher Agatston scores in CT data from 14,365 NLST participants.Two segmentation strategies were suggested by Kulathilake, L., G. R. Abdullah, Constantine, K. A. S. H. Ranathunga, N. A. et al. [17]. Frangi's Filter is used in the first technique, a unique vessel segmentation, while region-growing segmentation (also known as flood fill) is used in the second technique. This segmentation method has a fallout rate of 0.053, an average segmentation accuracy of 93.73%, and a sensitivity of 0.78.

A Hernandez-Gonzalez, Cervantes-Sanchez, F. Cruz-Aceves, I., Hernandez-Aguirre, M. A. Solorio-Meza, S. E. et al. [5] proposed a new automatic segmentation method based on multiscale analysis. In this method, Gaussian matched filters and multiscale Gabor filters were used for image domain and frequency domain respectively. These filters are used to improve contrast, background, noise suppression and homogenize the illumination of images. Both of the multiscale responses serve as the input layers for a four-layered perceptron network. This approach accomplished a Dice coefficient of 0.6857 and an accuracy of 0.9698. Kulathilake, Abdullah, Constanța, Ranathunga, N. A., et al. [17] proposed two segmentation techniques. The first technique involves a novel vessel segmentation using Frangi's filter, while the second technique is based on region-growing segmentation, also known as flood fill. The segmentation technique has a sensitivity of 0.78 and an average accuracy of 93.73%, with a 0.053 fallout rate. Hernandez-Gonzalez, Cervantes-Sanchez, Cruz-Aceves.

In order to automatically segment data in the image and frequency domains, Hernandez-Aguirre, Solorio-Meza, et al. proposed employing multiscale Gabor filters and Gaussian-matched filters. These filters improve an image's contrast, background, noise reduction, and illumination. A four-layered perceptron network's input layers are both the multiscale responses. The approach produced an accuracy of 0.9698 and a dice coefficient of 0.6857. A 3D-CNN was created by Candemir, White, Demirer, Gupta, Bigelow, Prevedello, and Erdal to identify pathogenic alterations in coronary arteries. Using a reference dataset of 247 atherosclerosis patients and 246 non-atherosclerosis patients, the system learns unequal properties for vessels with and without atherosclerosis. The study revealed a 58.8% positive predictive value, a 91.9% accuracy, a 93.3% specificity, and a 93.3% negative predictive value. The sensitivity was 68.69%, with a NPV of 96.1 percent at the artery/branch level, a threshold of 0.5 and an averaged region under the receiver operating characteristic curve of 0.91. CALD-Net, developed by Tianming Du et al., is an automated system using deep learning and convolutional neural networks, boasting a 0.88 recall rate.

Machine learning and deep learning techniques are combined with clustering and optimization techniques in the assessment of CAD and other cardiac diseases. Various deep learning models and different techniques applied by various researchers in CAD and related diseases, along with the accuracy achieved, are listed in Table 3.

Table 3. Various deep learning techniques and their achieved accuracies

Reference	Technique	Brief description	Achieved accuracy
[33]	LSTM	A deep recurrent neural network is utilized to predict patient outcomes after performing surgery via physiological parameters	91.6%
[34]	RNN	Recurrent neural network is used for heart failure prediction	88.3%
[35]	CNN	CNN is utilized for categorizing the subjects based on ECG beat classification	92.7%
[36]	Auto-encoder and neural network	The cardiac arrhythmia are assessed based on feature extraction and classification using neural network	97.99%
[37]	Deep belief network	ECG arrhythmia analysis has been carried out using Deep belief networks with fine tuning process	99.5%

5   Conclusion

CAD has become a common disease, and it is found to be the major cause of death in humans. General studies depict that a greater number of people are suffering from CAD almost everywhere in the world. CAD is becoming like a pandemic in developing countries. Most people are not aware of CAD, and it is becoming more adverse and tough to manage. By creating awareness about food habits, avoiding smoking and alcohol, taking proper rest, practicing meditation to overcome stress, and doing workouts to maintain good health, we may lower the risk rate of coronary artery disease. Deep learning has become a highly popular and highly prioritized technology in the field of artificial intelligence. Deep learning algorithms are utilized by many researchers to accurately detect CAD in its early stages, potentially saving lives. Various techniques and applications of deep learning have been discussed in this paper, as has the accuracy achieved by various works also presented.

References

1. Abdolmanafi, A., Duong, L., Ibrahim, R., Dahdah, N.: A deep learning-based model for characterization of atherosclerotic plaque in coronary arteries using optical coherence tomography images. Med. Phys. **48**(7), 3511–3524 (2021). https://doi.org/10.1002/mp.14909
2. Altan, G., Allahverdi, N., Kutlu, Y.: Diagnosis of coronary artery disease using deep belief networks. Eur. J. Eng. Nat. Sci. **2**(1), 29–36 (2017)
3. Al-Milli, N.: Backpropogation neural network for prediction of heart disease. J. Theor. Appl. Inf. Technol. **56**(1), 131–135 (2013)
4. Candemir, S., White, R.D., Demirer, M., Gupta, V., Bigelow, M.T., Prevedello, L.M., Erdal, B.S.: Automated coronary artery atherosclerosis detection and weakly supervised localization on coronary CT angiography with a deep 3-dimensional convolutional neural network. Comput. Med. Imaging Graphics **83**, 101721 (2020). https://doi.org/10.1016/j.compmedimag.2020.101721
5. Cervantes-Sanchez, F., Cruz-Aceves, I., Hernandez-Aguirre, A., Hernandez-Gonzalez, M.A., Solorio-Meza, S.E.: Automatic segmentation of coronary arteries in X-ray angiograms using multiscale analysis and artificial neural networks. Appl. Sci. **9**(24), 5507 (2019). https://doi.org/10.3390/app9245507
6. Chung, A.C.S.: Image segmentation methods for detecting blood vessels in angiography. In: 9th International Conference on Control, Automation, Robotics and Vision (2006). https://doi.org/10.1109/ICARCV.2006.345331
7. Dargan, S., Kumar, M., Ayyagari, M.R., Kumar, G.: A survey of deep learning and its applications: a new paradigm to machine learning. Arch. Comput. Methods Eng. **27**(4), 1071–1092 (2020). https://doi.org/10.1007/s11831-019-09344-w
8. Du, T., Liu, X., Zhang, H., Xu, B.: Real-time lesion detection of cardiac coronary artery using deep neural networks. In: Proceedings of 2018 6th IEEE International Conference on Network Infrastructure and Digital Content, pp. 150–154 (2018). https://doi.org/10.1109/ICNIDC.2018.8525673
9. Kaba, S., Haci, H., Isin, A., Ilhan, A., Conkbayir, C.: The application of deep learning for the segmentation and classification of coronary arteries. Diagnostics **13**, 2274 (2023). https://doi.org/10.3390/diagnostics13132274

10. Fischer, A.M., et al.: Accuracy of an artificial intelligence deep learning algorithm implementing a recurrent neural network with long short-term memory for the automated detection of calcified plaques from coronary computed tomography angiography. J. Thorac. Imaging **35**(May), 49–57 (2020). https://doi.org/10.1097/RTI.0000000000000491

11. Gessert, N., et al.: Automatic plaque detection in IVOCT pullbacks using convolutional neural networks. IEEE Trans. Med. Imaging **38**(2), 426–434 (2019). https://doi.org/10.1109/TMI.2018.2865659

12. Han, D., Liu, J., Sun, Z., Cui, Y., He, Y., Yang, Z.: Deep learning analysis in coronary computed tomographic angiography imaging for the assessment of patients with coronary artery stenosis. Comput. Methods Programs Biomed. **196**, 105651 (2020). https://doi.org/10.1016/j.cmpb.2020.105651

13. Heidari, A.A., Faris, H., Aljarah, I., Mirjalili, S.: An efficient hybrid multilayer perceptron neural network with grasshopper optimization. Soft. Comput. **23**(17), 7941–7958 (2019). https://doi.org/10.1007/s00500-018-3424-2

14. Jayasree, M., Rao, L. K.: A deep insight into deep learning architectures, algorithms and applications. In: Proceedings of the International Conference on Electronics and Renewable Systems (ICEARS), pp. 1134–1142 (2022) https://doi.org/10.1109/ICEARS53579.2022.9752225

15. Kim, J.K., Kang, S.: Neural network-based coronary heart disease risk prediction using feature correlation analysis. J. Healthc. Eng. **2017**, 2780501 (2017). https://doi.org/10.1155/2017/2780501

16. Koti, P., Dhavachelvan, P., Kalaipriyan, T., Arjunan, S., Uthayakumar, J., Sujatha, P.: Heart disease prediction using hybrid harmony search algorithm with Levi distribution. Int. J. Mech. Eng. Technol. **9**(1), 980–994 (2018)

17. Kulathilake, K.A.S.H., Ranathunga, L., Constantine, G.R., Abdullah, N.A.: A segmentation method for extraction of main arteries from Coronary Cine-Angiograms. In: 15th International Conference on Advances in ICT for Emerging Regions, ICTer 2015. Conference Proceedings, pp. 9–15 (2016). https://doi.org/10.1109/ICTER.2015.7377659

18. Lecun, Y., Bottou, L., Bengio, Y., Haffner, P.: Gradient-based learning applied to document recognition. In: Proceedings of the IEEE (1998). http://ieeexplore.ieee.org/document/726791/#full-text-section

19. Lee, S., et al.: Deep-learning-based coronary artery calcium detection from CT image. Sensors **21**(21), 1–15 (2021). https://doi.org/10.3390/s21217059

20. Liu, X., Du, J., Yang, J., Xiong, P., Liu, J., Lin, F.: Coronary artery fibrous plaque detection based on multi-scale convolutional neural networks. J. Sign. Proces. Syst. **92**(3), 325–333 (2020). https://doi.org/10.1007/s11265-019-01501-5

21. Nasr-Esfahani, E., et al.: Vessel extraction in X-ray angiograms using deep learning. In: Proceedings of the Annual International Conference of the IEEE Engineering in Medicine and Biology Society, EMBS, pp. 643–646 (2016). https://doi.org/10.1109/EMBC.2016.7590784

22. Nikan, S., Gwadry-Sridhar, F., Bauer, M.: Machine learning application to predict the risk of coronary artery atherosclerosis. In: Proceedings - 2016 International Conference on Computational Science and Computational Intelligence, CSCI, pp. 34–39 (2016). https://doi.org/10.1109/CSCI.2016.0014

23. Jegan, C.: Heart disease prediction system using supervised learning classifier. Bonfring Int. J. Software Eng. Soft Comput. **3**(1), 01–07 (2013). https://doi.org/10.9756/bijsesc.4336

24. Shadmi, R., Mazo, V., Bregman-Amitai, O., Elnekave, E.: Fully-convolutional deep-learning based system for coronary calcium score prediction from non-contrast chest CT. In: Proceedings - International Symposium on Biomedical Imaging, (ISBI), pp. 24–28 (2018). https://doi.org/10.1109/ISBI.2018.8363515

25. Shen, Y., Fang, Z., Gao, Y., Xiong, N., Zhong, C., Tang, X.: Coronary arteries segmentation based on 3D FCN with attention gate and level set function. IEEE Access **7**, 42826–42835 (2019). https://doi.org/10.1109/ACCESS.2019.2908039

26. Simson, W., Paschali, M., Navab, N., Zahnd, G.: Deep learning beamforming for sub-sampled ultrasound data. In: IEEE International Ultrasonics Symposium, IUS, pp. 1–4 (2018). https://doi.org/10.1109/ULTSYM.2018.8579818

27. Subhadra, K., Vikas, B.: Neural network based intelligent system for predicting heart disease. Int. J. Innov. Technol. Explor. Eng. **8**(5), 484–487 (2019)

28. Zotti, C., Luo, Z., Lalande, A., Jodoin, P.M.: Convolutional neural network with shape prior applied to cardiac MRI segmentation. IEEE J. Biomed. Health Inform. **23**(3), 1119–1128 (2019). https://doi.org/10.1109/JBHI.2018.2865450

29. Zreik, M., Van Hamersvelt, R.W., Wolterink, J.M., Leiner, T., Viergever, M.A., Išgum, I.: A recurrent CNN for automatic detection and classification of coronary artery plaque and stenosis in coronary CT angiography. IEEE Trans. Med. Imaging **38**(7), 1588–1598 (2019). https://doi.org/10.1109/TMI.2018.2883807

30. Gharleghi, R., et al.: Annotated computed tomography coronary angiogram images and associated data of normal and diseased arteries. Sci. Data **10**, 1–7 (2023). https://doi.org/10.1038/s41597-023-02016-2

31. Moona, J.H.: Automatic stenosis recognition from coronary angiography using convolutional neural networks. Comput. Methods Programs Biomed. Elsevier **198**, 1–11 (2021). https://doi.org/10.1016/j.cmpb.2020.105819

32. De Santis, D., et al.: Deep learning image reconstruction algorithm: impact on image quality in coronary computed tomography angiography. Radiol. Med. (Torino) **128**(4), 434–444 (2023). https://doi.org/10.1007/s11547-023-01607-8

33. Gopalswamy, S., Tighe, P.J., Rashidi, P.: Deep recurrent neural networks for predicting intra-operative and postoperative outcomes and trends. In: Proceedings. IEEE EMBS International Conference on Biomedical.& Health Informatics, pp. 361–364 (2017)

34. Choi, E., Schuetz, A., Stewart, W.F., Sun, J.: Using recurrent neural network models for early detection of heart failure onset. J. Amer. Med. Inform. Assoc. **24**(2), 361–370 (2016)

35. Zubair, M., Kim, J. Yoon, C.: An automated ECG beat classification system using convolutional neural networks. In: Proceedings of the IEEE 6th International Conference on IT Convergence Security, pp. 1–5 (2016)

36. Luo, K., Li, J., Wang, Z., Cuschieri, A.: Patient-specific deep architectural model for ECG classification. J. Healthc. Eng. **2017**, 1–13 (2017). https://doi.org/10.1155/2017/4108720

37. Wu, Z., et al., A novel features learning method for ECG arrhythmias using deep belief networks. In: Proceedings of the 6th International Conference on Digit Home, pp. 192–196 (2016)

Author Index

© The Editor(s) (if applicable) and The Author(s), under exclusive license
to Springer Nature Switzerland AG 2024
S. Satheeskumaran et al. (Eds.): ICICSD 2023, CCIS 2122, pp. 213–214, 2024.
https://doi.org/10.1007/978-3-031-61298-5

Printed in the United States
by Baker & Taylor Publisher Services